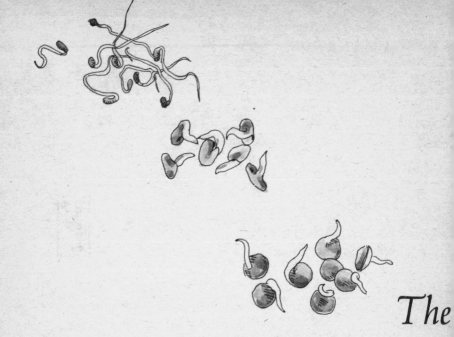

The

ILLUSTRATIONS BY LORRAINE BODGER

Beansprout Book

By Gay Courter

SIMON AND SCHUSTER

NEW YORK

ISBN 0-671-21596-5
ISBN 0-671-22947-8 Pbk.
Library of Congress Catalog Card Number 73-8223
Designed by Irving Perkins
Manufactured in the United States of America

4 5 6 7 8 9 10 11 12 13
      6 7 8 9 10 11 12 13 Pbk.

FOR PHILIP

My most patient and loving sprout taster

ACKNOWLEDGMENTS

*With special thanks for information and ideas,
help and encouragement to:*

Leesa Barenboim
Sandra Branch
Dorothy and Daniel Courter
The Aaron Everetts
Julie Houston
Marylee Kendrick
Cyrus and Charlie Porter
Frederic Beale Sadtler
John and Carrie Staples
Rosalind Stevens
Pat Tsien
Elsie and Leonard Weisman
Robin Weisman

Contents

And God said, Behold, I have given you every herb bearing seed, which is upon the face of all the earth, and every tree, in the which is the fruit of a tree yielding seed; to you it shall be for meat.

—GENESIS 1:29

Introduction

It was Thanksgiving at Antioch College in Yellow Springs, Ohio. A college professor invited several students who hadn't gone home for the holiday to a feast at his home. The turkey and trimmings were traditional, but he served an unusual vegetable—one with the crispest, most memorable taste. He called it "sprouts" and said he grew them on his kitchen counter. He offered to teach me how to grow them, but I was, well, disinterested. I ate my meals in the cafeteria, and my interest in food preparation was minimal. Besides, no one grew vegetables in the dorm.

Several years later, when I began cooking for myself, I regretted that I never learned how to grow a sprout. When passing the bean counter in a supermarket I would wonder, "Is this the bean that made that vegetable? How much soil would I need? How many months would it take to grow them?" In Chinese restaurants I would separate each vegetable in a dish and study it. "Is this a sprout? Where do they come from?" I tried canned bean sprouts—the name was right, but the taste insipid, the color and texture disappointing. I questioned cooks I knew and respected, but no one had any clues to this fabulous food. I wrote Antioch, but the professor had departed. I described them to my husband, but he thought I was imagining that something like sprouts could be grown at home.

Then one day I ventured into a new health-food store looking for stone-ground flours. A small sign above a counter read, "Seeds For Sprouting." Eureka! Excitedly I asked the clerk if those were the incredible little beans

9

that taste so good, and that you can grow in your own kitchen. They were indeed.

I went home with five types of beans, some sheets of directions describing various methods of sprouting, and a clever-looking commercial sprouting container. The directions said it would take only three to five days before I could harvest my crop. That seemed impossible. And all I needed was water—no soil or nutrients.

I followed the directions, using the reliable mung bean. Every few hours the beans swelled, changed and grew. After a three-day vigil beside the sprouter, they were ready to harvest. We ate sprouts by the handful raw, tossed them into the soup, then the salad, and finally sautéed them—all at the same meal!

Since then I often serve sprouts in one form or another. When guests ask what they are eating and where they can get it, I show them my kitchen garden. This is partially to impress them and partially to proselytize them. I usually have to type out pages of notes to get novices started, because I found none of the commercial directions as satisfactory as the methods I finally worked out for myself.

Now my husband's eighty-year-old grandmother and my teen-age sister sprout, a bachelor friend sprouts to impress his girl friends, a veteran hiker sprouts in his backpack, and a sailor friend sprouts on his yacht. I suppose it's because I've sprouted and served so many pounds of beans and seeds that I'm called on for expert advice so often, and I haven't decided if it's flattering or not to be introduced as "the beansprout lady." But even though I enjoy sharing my sprouting experiences with others, I am tired of typing long pages of directions, giving recipes over the phone, and answering distress calls.

So here it is: everything I know about sprouts and some terrific ways to use them up. It's easy, fast, delicious, nutritious and cheap. You too can become the sprout lady or gentleman on your block. Happy sprouting!

Why Sprout?

SPROUT FOR ECOLOGY

Sprouting your own food gives you complete control over the quality of what you eat. You know exactly what seeds or beans you are using and where they have come from. You know they have been properly rinsed and cleaned. You know which water supply has nourished the sprouts. You know that the containers have been cleaned and you know exactly whose hands have picked them over.

There have been no pesticides, preservatives or additives used on your home-grown sprouts. No trucker, clerk or shopper has handled them. You know the exact day the sprouts started to grow and can calculate when they will be at their optimum and when they will begin to lose their nutritional potency.

Unless you spend months on elaborate gardening, or manage to raise your own meat supply, or process your own dairy products, you will never again have so much involvement with and influence over what you are eating. Now you can have a complete, wholesome and natural food, without any unknown quantities.

Another factor to consider when you sprout at home is the great savings in natural and human resources. Compare growing the sprout on your kitchen counter to purchasing a canned or frozen vegetable—spinach, for instance—at the supermarket. Farmers prepared the land, planted the seed, where it sprouted underground, supervised the growth of the plant, watered, sprayed and worried over the crop until men and machines harvested it. Then the spinach was transported to a factory, processed again into its packaged state. Finally it traveled, possibly thousands of miles, to be warehoused and transported again to the market. Now you purchase the spinach, transport it home, unpack and store it. Later you unwrap, cook and serve it to your family. How many man hours went into that package of spinach? How much electricity, diesel fuel and gasoline were consumed by the machines that planted, harvested, packaged and transported it? How much waste was created by the manufacturing of the cardboard package or the tin can? How much pollution by the burning of the fuels? How much water was used up along the way?

Growing some food at home is a real way that you can become involved with the basics of life, eliminating the middlemen who sap the power you could have over your own biological needs.

SPROUT FOR HEALTH

The sprouted seed is known to be one of the most nutritious foods available. Something wondrous happens as the little seed swells and sprouts, releasing dormant energy and vitamins. Each seed contains an embryo which is a miniature of what the plant will be. This endosperm, or nutritive tissue, contains a supply of stored carbohydrates, oils and proteins to nourish the seed. When environmental conditions are suitable, the seed will germinate, living on its own supply of food until it has grown sufficiently to have roots which can absorb nutrients from the soil, and to have leaves that open up to the sun. Sprouts are harvested at the point where the seed is manufacturing all those life-giving nutrients, but before the embryo has a chance to completely consume the nourishment and deplete its vitamin-rich storehouse.

Recently sprouts have been rediscovered by nutritionists, who have found them rich in almost every important vitamin and mineral while also containing enough proteins to be classed as a "complete food." Many of the sprout proteins are predigested, for they are converted to amino acids during the sprouting process. The starches are also converted to simple sugars requiring little digestive breakdown, so they enter the bloodstream rapidly and are classed as a quick-energy food.

Sprouts also contain enzymes, the complex catalysts controlling many of the chemical reactions that take place in our bodies. We manufacture fewer and fewer enzymes as we age, and since foods cooked at temperatures greater than 140° F. kill them, our stock of enzymes must be replenished by eating fresh produce. This is another reason for consuming home-grown sprouts.

The high amounts of Vitamin C contained in the tiny sprout is truly amazing. Many sprouts contain as much (or more) Vitamin C as is found in an equal quantity of citrus fruit juices. Early investigations in the value of sprouts were conducted by Dr. Cyrus French during World War I. He selected troops suffering from scurvy and divided them into two groups. One group received four ounces of lemon juice a day, the others were given four ounces of sprouted beans. Within a month, over 70 percent of the bean eaters were free from scurvy symptoms, compared to only 53 percent of the lemon-juice takers.

A study in England concerned African laborers who developed scurvy. The Africans were used to drinking large quantities of beer in their homeland. This native beer, made from sprouted millet, was rich in vitamin C. But they soon showed symptoms of scurvy after drinking the European beer, which offered no such enrichment.

Many studies of sprouted grains have been conducted with animals. Research with cattle at the Agricultural Experiment Station in Beltsville, Maryland, has shown that problems of infertility in cows can successfully be treated by including large amounts of sprouted oats in their diet.

As the seeds begin to sprout, their vitamin content also begins to grow. The first early shoots of soybeans (per 100 grams of seed) contained only 108 milligrams of vitamin C in one study conducted at the University of

Pennsylvania. But after 72 hours the vitamin C content had soared to 706 milligrams, an increase of almost 700 percent! Dried peas increased from 0 milligrams of vitamin C to 69 milligrams after 48 hours, and 86 milligrams after 96 hours. Similar comparisons can be made for all the vitamins, most of which increase greatly in potency during the sprouting process. In some cases it's the B vitamins or vitamins A or E that make the highest increases. The rate of vitamin increase during sprouting, and the ultimate decrease of nutritional value when the endosperm is consumed, varies with each type of bean or seed used.

I'm convinced by numerous studies that sprouts do indeed contain a varied and powerful battery of nutrients, rivaling citrus fruits in vitamin C and beef in protein, and surpassing almost any other known food source in completeness. By sprouting seeds and beans yourself, *you* control their potential for health by eating them at their prime to get the most food value for your time and money.

SPROUT FOR TASTE

There are many foods and products which are supposed to be good for you, but that doesn't mean you'll eat them. You may be induced to try such health foods as brewer's yeast, creamed papaya or kelp, but if they don't taste terrific, you're not likely to continue to consume them on a regular basis.

Luckily, sprouts have almost universal appeal. There are so many varieties, each with its own texture and taste, that there will be at least one sprout to please everyone. The classic mung bean, used so often in Chinese cooking, is crunchy and mild. If you eat and enjoy any raw greens—lettuce, celery, or green pepper—you will enjoy the mung-bean sprout.

Some raw sprouts, such as soy, have the distinct fresh taste of just-picked garden peas, which delights some but repels others. When you steam that same raw sprout for a few minutes, the taste changes to a nutty and crisp flavor that everyone seems to enjoy. Other sprouts, such as wheat and alfalfa, are surprisingly sweet, and they satisfy a craving for something sugary without adding calories. Rye sprouts are often mistaken for wild rice when served in soups and rice combinations. Gourmets looking for exotic flavors will find that fenugreek sprouts suggest Indian foods and go well with curries, while raw lentil sprouts add a peppery tang to salads.

The flavors of sprouts change considerably in combination with one another, when cooked in sauces and gravies, and when used in different recipes. Even the fussiest eaters and those on the blandest diets will find some sprout or sprout dish appealing. People who regularly shun beans because of their gas-producing qualities happily report that the sprouting process manages to liberate the gas before it enters the human digestive system.

Yes, sprouts *are* good for you, with the added bonus of being an absolutely delicious vegetable with endless flavor possibilities.

SPROUT TO SAVE MONEY

Another very important reason to perfect your sprouting technique and to sprout on a regular basis is the amazing economy of growing the sprouts. It is so inexpensive to feed sprouts to my family and friends that I sometimes wonder why I haven't been sprouting all my life. Why didn't anyone tell me about this before? Why aren't sprout materials and beans stocked in every supermarket? Why aren't they nationally advertised? Why isn't there a sprout industry?

The answer to these questions is simple: Sprouts are so inexpensive to begin with and a little goes such a long, long way, that it just wouldn't pay big business to become involved in the marketing and distribution of sprouts. If Americans became used to growing this scrumptious vegetable, the bottom might fall out of the canned and frozen vegetable market.

The most expensive sprouts (such as alfalfa, black radish, mustard) purchased in small quantities may run as high as $3.30 per pound. But remember that one pound of dry beans will make about eight pounds of sprouted seed. And one pound of freshly sprouted seed may serve six to eight people. Looking at it that way, each serving costs between $.05 and $.07. Compare that to your favorite frozen vegetable, which will cost between $.10 and $.15 per serving. Most sprouts sell for considerably less. The favorite mung bean retails for between $.90 and $1.30 per pound, or a cost of about $.03 per serving. And the most nutritious of all the sprouts— the soybean sprout—usually sells for less than $.70 a pound, costing but $.01 or $.02 a serving.

If you really get into sprouting, you'll be purchasing your seeds in larger quantities, as they store easily and have a long shelf life. I've found an excellent-quality soybean that costs only $1.30 for ten pounds. Now my per serving cost is a ridiculous $.002!

During extremely difficult times a family of seven survived a Utah winter by eating nothing but sprouts for six full months. They spent a total of $52.50 for their food at a cost of $.015 per person per meal. No one suffered any ill effects, and during those six months the family was free from colds and disease. This is certainly an extreme example, but it points up the fact that if you regularly include some form of sprouts in your menu planning, you can't help but reduce the family food budget, without anyone's suffering hunger, boredom or malnutrition.

SPROUT TO SAVE CALORIES

Dieters rejoice at the thought of sprouts. One fully packed cup of the most common sprouts (mung, alfalfa, radish) contains only about 16 calories. They can be eaten raw or cooked simply, and they taste delicious without the need for rich sauces, gravies or starches. Sprouts are an easy way to satisfy the dieter's need for protein without consuming the high calories usually associated with most protein-rich foods. Soybeans, peas and lentils (which are extremely high in protein) do contain more calories than some of the other sprouts— about 65 calories per cup. But the yield of protein, weight for weight, is approximately twice that of meat and, in the case of soybeans, four times that of eggs.

Sprout flavors and textures are varied enough to help keep even the most bored dieter on the straight and narrow. A handful of wheat or rye sprouts adds up to only about 8 calories and is a delightful sweet treat. Because the sprouts are a well-balanced food in terms of vitamins, minerals and proteins, they can help the long-term dieter and those on weight-maintenance programs feel confident that their nutrition won't suffer for the sake of their figures.

SPROUT FOR FUN

By now you are no doubt convinced that sprouts will erase your suspicions about where your food comes from and will help you do an ecological service to your country. You can also look forward to sprouts helping to keep you healthy and thin, and to make you wealthy. But have you ever thought of the simple satisfaction you will get from watching the little things grow? I had always wanted my own vegetable garden minus the five Ws of weeding, watering, waiting, worrying, and WORK! My bean-sprout patch grows to maturity in three to five days and has never failed me yet. My sprouts are a constant source of wonder and satisfaction.

Children, too, become fascinated with the sprouting process. One of the most creative teachers I know started a sprouting project with her first-graders. She had worked on other indoor gardening units with her class, but the success rate was disappointing. No one was able to tend the plants over long weekends and vacations and their slow growth caused some of the youngsters to lose interest in their project.

She started her class on sprouts one Monday, by examining the seeds and talking about plants and how they grew. Each child soaked his beans overnight. Twice a day each child went to the sink and rinsed his beans. The class discussed the sprouting process day by day from the emergence of the shoot to a debate on when they would be good to eat. Finally on Friday, they had a sprout-cooking-and-eating party. Extra sprouts were sent home for the children's families to enjoy.

Even the youngest child can have a little sprout garden at home near the kitchen sink. It's a simple responsibility, one that's easy for a child to meet. Caring for his own sprout patch allows the child to take a creative part in providing food for his whole family. This gives him a sense of worth and achievement, for in a very real sense he is making a contribution to the physical welfare of others. I've yet to meet the child who refused to eat the vegetable he had raised proudly all by himself.

Who Sprouts?

I only discovered sprouts quite recently, but there is a persistent and dedicated society of people who have been sprouting for many years. As I began my sprout research, I discovered a strong sprouter's underground led by families of Chinese extraction who regularly sprout the mung and soy beans so common to their gastronomic culture. Oriental specialty stores and restaurants often carry freshly grown sprouts.

Health-food aficionados have long been disseminating information on the value of sprouts in the diet. Today almost every town has an outlet for health-food products. Tucked away on the shelves in these counterculture markets are nondescript bags containing the dried seeds and beans used for sprouting.

Early advocates of the beansprout found their way into the Bible. The "pulses" mentioned in Chapter 1 of the Book of Daniel (verses 12–16) are said to be the first reference to sprouts in the West.

More modern sprouters included Sir Francis Chichester, who sprouted on his solo round-the-world sailing in the *Gypsy Moth IV*.

Sprouting is taught as a survival tool, and no well-equipped bomb shelter or life raft should be without some food that will sprout.

Some devotees claim sprouts are a potent healer. I don't think they are a panacea, but significant work has shown their usefulness in treating beri-beri, scurvy, malnutrition, slow bone healing, and infertility.

Submarine crews often find sprouts on the menu, for it is the one fresh vegetable they can harvest while underwater for long periods of time.

I haven't heard of any astronauts sprouting as yet, but certainly long space flights and orbiting space laboratories will be equipped with sprouting apparatus.

A hiker can easily sprout in his knapsack, so there is no reason why a student can't sprout a small amount of beans in a plastic bag tucked in his jacket pocket or bicycle basket.

Farmers sprout grains to increase the health and virility of their livestock. Maybe they secretly take some home for dinner as well?

Truckers and salesmen might consider sprouting on the road. I know many secretaries and office workers who sprout right at their desks, tossing the finished sprouts into bouillon, sandwiches, yogurt or salads for a fast high-energy meal.

There's no place too small or remote for sprouting. If you can reach a supply of water twice daily, and if the temperature is within the range of comfort for human habitation, seeds will sprout successfully.

Sprout a bean today, no matter who you are or where you are! Some of the most creative crops have been raised by a hardy breed of flagpole sitters. And it's even easier to try sprouting in the comfort of your own home.

You Sprout!

SPROUTING EQUIPMENT

You can spend between $5.00 and $25.00 to purchase sprouting apparatus that will successfully sprout most beans, but there are probably a dozen containers in your kitchen that will work equally well, as I found when testing seeds and beans for sproutability. I used every conceivable pot, bowl, crock, opaque apothecary jar and storage container in the house, and all were a success.

My basic sprouting system requires any receptacle large enough to hold the finished sprouts, but this container must *not* be transparent, wooden or metallic. If you wish to sprout in glazed pottery, it *must* be high-fired stoneware and never low-fired earthenware, which may contain toxic lead sulfate in its glaze. Low-fired earthenware is usually inexpensive, porous, and often of Mexican origin.

Containers of plastic, china, enamel and unglazed pottery are excellent choices. Since the sprouting container must be kept covered, anything that comes with its own lid is a good choice, including bean pots, crocks, canisters, coffeepots, fondue pots, cookie jars, chafing dishes, large plastic storage containers (Tupperware, Rubbermaid), and plastic ice-cream boxes.

Any plastic or china mixing or serving bowl works if a dinner plate is placed over the top to prevent the sprouts from drying out.

A container with a wide diameter is best when sprouting larger quantities of sprouts. Don't use tall, narrow or small-mouthed containers. Try to have your sprouts in as few layers as possible, for even circulation of moisture and air. If some sprouts on the bottom are weighed down

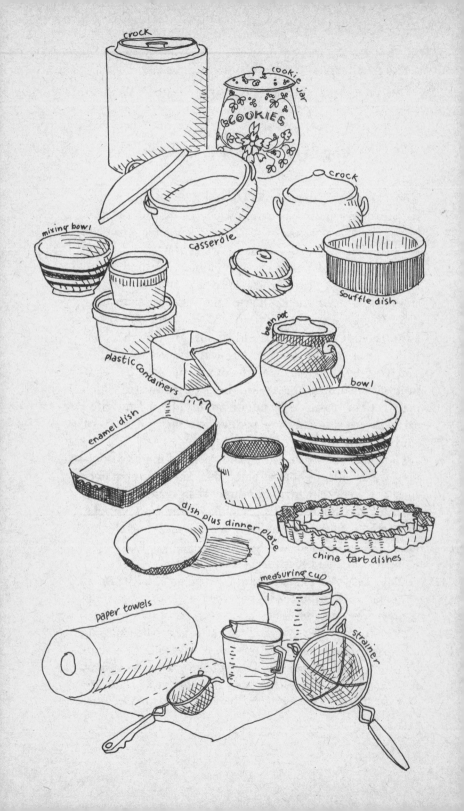

by the others on top, the pressure will increase the chances for problems of rot and spoilage to develop.

Once you have selected your container, you will need only a few other pieces of equipment:

- Measuring cups (2- and 4-cup)
- Paper towels
- A large wire mesh strainer

A measuring cup is used to measure the initial amount of dried seeds to be used and is also convenient for soaking the seeds. Since I usually choose to sprout ¼ cup of dried seeds at a time, I soak them in a 2-cup measure.

Moist paper towels are placed on top of the sprouts to help provide the correct percentage of moisture in the sprouting atmosphere. I prefer white towels for this purpose. If the towels sour, replace them immediately. Sometimes a dry towel at the bottom of the sprouter will help absorb moisture if you are having a chronic water-accumulation problem.

A strainer will facilitate rinsing and draining your sprouts at regular intervals.

SPROUT THEORY

There are three basic factors that control the sprouting of a seed:

1. The amount of moisture
2. The proper temperature
3. The circulation of air

When dried seeds or beans become moist, they wake from their dormant state and begin their irreversible growth process. During this process of germination, chemical changes begin to take place; carbon dioxide, other gases and heat are released. These gases and residues create wastes that will accumulate if not permitted to dissipate. One of the most important steps in the sprouting process is to keep removing these wastes by rinsing the sprouts with fresh water to prevent the crop from souring and spoiling. Cool water ventilates the sprouts and prevents their overheating and destruction.

While sprouts demand a constant supply of moisture to grow, they cannot be allowed to sit in water or they will rot. It's not difficult to sprout if you make certain that the sprouts are always moist, but never left standing in even the smallest puddle of water.

Sprouts grow fastest in warm temperatures, free from drafts and away from direct heat. In cold weather, soaking times may be increased a few hours and the sprouts can be rinsed in slightly warmer water. Between 75° and 85° F. is the ideal sprouting temperature.

Air must be allowed to circulate in the sprouting container. There should always be about one-third of the sprouter left empty for air circulation. Remember that sprouts expand, and provide plenty of growing room for them.

Simply stated, these are the basic rules of sprouting:

1. Keep your sprouts moist, but never wet.
2. Keep your sprouts warm.
3. Rinse your sprouts as often as possible.
4. Give sprouts room to breathe.
5. Don't cram your sprouts in too small a container.

MY BASIC SPROUTING METHOD

Select the seed or bean you wish to sprout. I highly recommend that neophytes try the mung bean first, for both its reliability and universal taste appeal. A few dried beans go a long, long way. I usually sprout ¼ cup dried beans at a time. Depending on the variety, this will yield approximately 2 cups at maturity, or about 4 servings.

An easy way to figure the yield is: 1 ounce of dried

seed will equal 1 cup mature sprouts. Thus, ½ cup, or 4 ounces, will equal about 4 cups sprouts.

A good timetable is to start soaking beans after dinner one night. Place them in the sprouter the next morning. Rinse every evening and every morning. Most sprouts will be ready to eat the evening of the third, fourth, or fifth day. Step-by-step, here is the procedure:

DAY 1

- Place the beans in the strainer and rinse for a few minutes in cool water to clean off any impurities, dust or pesticides.
- Soak the beans in at least 4 times as much warm (80° F) water as beans.
    ¼ cup beans to 1 cup water
    ½ cup beans to 2 cups water
    1 cup beans to 4 cups water, etc.
- Let the soaking beans stand in a warm place for 8 hours or overnight. In cooler weather, it is wise to let the beans soak as long as 12 to 16 hours.
- The soak water, loaded with water-soluble vitamins and minerals, has a mild sprouty flavor. Many people drink it, or add it to soups, or use it as a cooking liquid with sprouts or other vegetables. If you're hesitant to consume it yourself, your house plants may enjoy the extra burst of nutrition when they're watered!

DAY 2

The beans will have at least doubled in bulk during the overnight soaking process. You may notice some small gas bubbles at the surface of the soak water, indicating that the sprouts are already releasing energy and heat in the early germination process.

Rinse and soak
8 hours or more

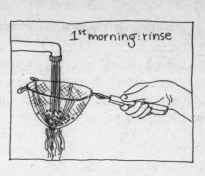

1st morning: rinse

Place in container
and spread evenly

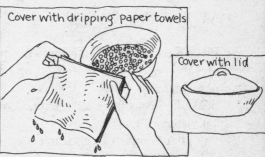

Cover with dripping paper towels

Cover with lid

1st evening: fill with
1 inch of water

Drain sprouts by
tilting

OR

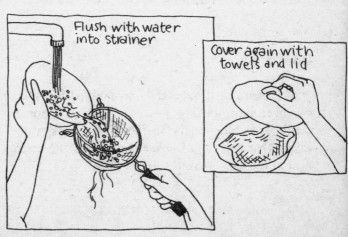

Flush with water
into strainer

Cover again with
towels and lid

- Place the beans back in the strainer and rinse in cool water.
- If a seed refuses to sprout, it just sits there, softens and rots, giving off an unpleasant, unappetizing odor. Now is the time to remove the seeds that are not germinating. Floating seeds indicate sterility, as do any with cracked or broken hulls.
- Place the selected, swelled seeds in your container. Spread them evenly along the bottom.
- Soak four sheets of paper toweling in warm water and carefully lay them dripping wet on top of the beans. Cover and set the sprouter aside. Let stand at room temperature in a convenient place, near the kitchen sink.
- In four hours sprinkle the paper towels with a few drops of water. This will help the seeds sprout faster, but this step can be eliminated if you don't have the time during the middle of the day.
- In the evening, set aside paper towels and let water run into the container to about one inch above the sprouts. If you have time, let the sprouts stand in the water for 5 to 15 minutes. This helps them plump up and grow faster.
- Cover and tilt the container, letting the water flow out from the tiny opening between the lid and rim. It will be necessary to drain seeds that are smaller than mung beans into a strainer first and then tap them back into the sprouting container.
- Some seeds may cling to the sides of the sprouter. Flush them with water and drain into the strainer. Repeat until all beans are in the strainer.
- Drain completely!
- Gently stir the beans with your hands or a wooden spoon so that the bottom sprouts move toward the top to encourage even sprouting. Push the seeds that cling to the sides down to the bottom.
- Moisten towels with water. Spread over the top of the sprouts. Cover and set aside.

DAY 3

Already you will notice considerable growth in your sprouts. The outer shell or husk will be falling off.
• Fill the container with water, drain well. Moisten towels, cover and set aside as you did on Day 2.
• You may choose to remove the husks that float to the surface when the sprouter is filled with water.
• Moisten towels in the middle of the day, if you have a chance.
• Repeat rinsing process in the evening.

DAY 4, DAY 5, AND DAY 6

Repeat the procedure for Day 3, if sprouts are not finished (see Harvesting, below).

This procedure may sound tedious. But as soon as you have learned the steps, you will find that you spend less than five minutes' total time each day tending your sprout garden.

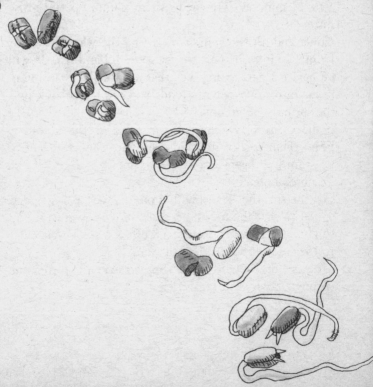

HARVESTING

When are they ready to eat? There are no hard-and-fast rules about when to harvest a sprout. Each variety tastes best at a different length, and three "experts" will give you three different perfect harvesting lengths, sometimes varying by several inches in their recommendations. The flavor varies at each point along the way in many sprouts, so settle the controversy by pleasing your own palate. As you sprout, keep tasting until you find the optimum day.

Some sprouts, such as sunflower, fenugreek and sesame, become decidedly bitter if left to sprout too long. Wheat and rye become sweeter as time goes on. Mung beans don't seem to change much in flavor, but the Chinese prefer to harvest them when they are quite long—at least two to three inches.

If sprouts are left in the sun between three and twelve hours before harvesting, the chlorophyll content will increase. All sprouts contain some percentage of chlorophyll, the substance that gives plants their green color and the energy to carry on the process of photosynthesis. Chlorophyll is chemically related to hemoglobin in the human blood, the only difference being that chlorophyll contains magnesium, while the blood contains iron. Many nutritionists recommend that we consume more chlorophyll to help build our blood, aid digestion, stimulate tissue growth and help freshen the breath.

The Chinese traditionally bleach the sprouts to remove all traces of chlorophyll. While the white sprout may add a decorative touch to a dish, the whiteness does not indicate purity. The bleached sprouts are definitely less nutritious.

If you want to include more chlorophyll in your diet, the sprouts should be given a few hours of sunlight before they are eaten. Take the sprouts and spread them in a long nonmetallic pan or tray in a single layer. Dampen them slightly, cover with clear plastic wrap to hold

29

moisture in and prevent drying out, and set in sunny place. The green should develop in a few hours. Too much light will toughen most sprouts. Alfalfa, clover, and mustard are delicious with their first tiny green leaves.

For a guide to the prime moment for harvesting each particular seed or bean, check my recommendations under each specific variety given in the Dictionary on page 44.

PREVENTING CROP FAILURE

You will first realize that your sprouts are in trouble if they aren't sprouting and/or if they emit a foul odor. These problems are most common with larger beans such as soy, garbanzo, kidney, pinto, etc. These are the major causes for failures in sprout crops:

1. Seeds do not germinate properly.
2. Seeds rot in too much water.
3. The sprouts dry and wilt in too little moisture.
4. The temperature is too cold.
5. The sprouts are exposed to too much light.
6. The sprouts are in a metallic container.
7. The sprouts are too crowded together and are deprived of enough air circulation.

SOLUTIONS

1. Try to buy seeds that are packaged especially for sprouting. They should be young seeds which have been stored properly.
2. Carefully remove any seeds that refuse to show any signs of sprouting. They may just soften into a

mush, remain a hard kernel, crack open and never show a sign of sprouting, or give off an odor.

3. Make certain that no beans are standing in water. Drain and turn regularly.

4. Sprout in fewer batches in larger containers, rather than crowding too many beans into small ones.

sprouts too crowded   sprouts not crowded   large container is even better

5. Keep sprouts at a temperature between 70° and 85° F.

6. Never use transparent containers if sprouts will be exposed to light. Never use metallic containers.

7. If rotting problems persist, place dry toweling at the bottom of the container to absorb any water that might seep down and accumulate at the bottom. Change the bottom towels every rinsing period.

8. Change the moist towels covering the top of the sprouts more frequently.

9. Sometimes the hulls that fall off and accumulate at the bottom and sides of the container may harbor bacteria. This is sometimes a problem with the larger beans. Since these tend to have hulls which are difficult to digest anyway, you may choose to remove them. Stir the beans gently while they are covered with water during the rinsing period and the loose hulls will float to the top. Skim them and throw them away.

10. If seeds still fail to sprout properly you can rinse the beans in an antibacterial solution of 1 teaspoon Clorox to 3 gallons of water once a day. This is used by some commercial sprouting establishments, but diligent rinsing with fresh water should work equally well.

11. Harvest beans when they are ready for eating. If

they won't be used right away, rinse them, wrap them in moist paper towels, place them in a plastic bag and refrigerate. This will keep them fresh for at least five days. Left at room temperature, they will deteriorate rapidly.

12. Keep sprouts away from excess heat. Don't place the sprouter on a stove, in an oven, over a pilot light or radiator.

13. Keep sprouts away from toxic fumes.

14. A few bad seeds won't ruin a whole crop. If you carefully remove any beans showing signs of rot or mold and freshen the others with water, they will be perfectly safe to eat.

THE PORTABLE KNAPSACK SPROUTING METHOD

Here is a way to carry your sprouts wherever you go, by sprouting in a lightweight, flexible plastic bag. You need only rinse them with water morning and evening to have fresh vegetables for your meals away from home.

You will need the following materials:

1. Seeds for sprouting—mung, lentil, rye, wheat and soy work best. Smaller seeds may slip through the holes you will make in the inner bag (see step 3, below, and step 2 under Directions, below), as the holes stretch during rinsing.

2. A heavy plastic bag. I use a yellow bag, the kind that holds garbage scraps, or a bread wrapper. Another possibility might be a packaging bag, the kind of heavy plastic bag used for new sweaters, blankets or pillows. A small waste-can liner can be used as is, or cut down to size.

3. A second lightweight plastic bag. I use the kind that come in rolls and are meant for food storage (11½" x 13").

4. A few paper towels or a clean white rag.

5. Twister seals or rubber bands.

DIRECTIONS FOR KNAPSACK SPROUTING:

1. Soak the beans or seeds in the heavy bag in 4 times the amount of water as there are beans for 8 hours or overnight. Seal bag with twister to keep the water in.
2. In the morning, take the lighter bag and punch small holes over about ⅓ of the bag's lower surface, 1 inch apart, with a sharp pointed object.

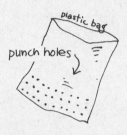

3. Pour the soaking beans through the sieved bag and shake, or hang the bag on a tree limb until all the water is drained out.

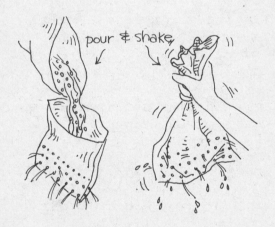

4. Soak the towels or cloth in water, squeeze out gently and tuck around the beans in the sieved bag.
5. Place the sieved bag containing the beans and moist toweling into the heavy bag. Twist the tops of the

two bags together and fasten with tie or rubber band to hold in the moisture.

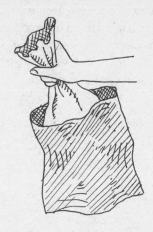

6. Tuck the two-bag sprouting system into your knapsack, glove compartment, suitcase, pocket, balloon basket, parachute, bicycle pouch or desk drawer and go on your way.
7. When you stop for the evening, take the sprouts to a water supply. Remove the inner sieved bag containing the sprouts. Flush with water for a few minutes. Drain completely. Moisten towels or cloth and tuck around sprouts. Place the sieved bag back inside the protective heavier bag. Seal with twister. Repeat in the morning and go on your way again.
8. Repeat rinsing and draining twice a day until you harvest your crop.

SPROUTING LARGE QUANTITIES

If you are sprouting for your family of twelve, or a commune or corporation, or if you plan to go into the business of selling sprouts for fun and profit, you will want to sprout in large quantities.

The following method is used by many Chinese restaurants for sprouting mung beans.

1. Fill a large 15-gallon crock or plastic garbage pail approximately ⅕ full of beans.
2. Fill the crock to the brim with warm (80° F) water.
3. Let stand overnight.
4. Drain. Cover with wet turkish towel and tight lid.
5. Flush with water and drain 2 to 3 times daily. Moisten the towel each time. (Some people do not bother to fill the crock with water each time. They stir the beans gently, changing their position in the crock, moisten the top towel more often, and rinse with water only if the beans seem to be drying out.)
6. Repeat daily. Sprouts should be ready to eat or refrigerate on the 4th or 5th day.

Mung beans work best in quantity, but rye, alfalfa, wheat and lentils also have a high rate of success. Experts sprout soy successfully in quantity. The beans must be of the highest quality and of the previous year's crop. Soybeans must be rinsed and drained faithfully to ensure success.

Quantity apparatus for sprouting may be improvised out of clam pots, beer kegs, and other containers with bottom spigots to facilitate draining. The whole contraption can be placed in a shower stall during the sprouting process. The shower spray provides the water source for rinsing, the spigot is turned on, and the water escapes directly into the shower drain.

OTHER SPROUTING METHODS

There are dozens of other ways to sprout successfully. The basic sprout method that I have outlined, improvising with household equipment, has worked best for me. I recommend it highly for all seeds and beans. The knapsack method is a unique solution for a lightweight portable garden, and there are many ways to improvise sprouters for extremely large quantities. Some people feel more secure with store-bought sprouters. There are several excellent models available, from unglazed pottery sprouters to a three-tiered Plexiglas showpiece. Here are some other methods that are commonly used for sprouting.

1. *Jar or bottle method.* Use a bottle or jar with a wide mouth. Place drained seeds at the bottom. Cover the mouth with cheesecloth or a thin wire screen, fastening securely with string, rubber band, or mason jar ring. Each time the sprouts are rinsed, invert the jar for a few minutes and drain completely. If jar is glass, store in cupboard out of sun or direct light.

2. *Coffee percolator.* This is good for small seeds. Scour well and place drained seeds in strainer and lower into pot. Cover. Rinse and drain frequently.

3. *Tea strainer.* Useful for sprouting individual servings of small seeds. Set strainer containing drained seeds into teacup or small bowl, cover with paper towels and then place saucer on cup.

4. *Colander.* Useful for larger seeds. Place drained beans in colander and set in large bowl. Cover with wet towels and dinner plate.

5. *Unglazed flowerpot or flowerpot saucer.* Soak pot by submerging in water for a few minutes. Plug drainage hole. Place seeds on bottom, cover with saucer and place in a shallow pan of water.

There are infinite variations of the same ideas. They all have the same purpose—to allow sprouts to keep moist, but not wet, and to facilitate rinsing and draining.

What Sprouts?

BUYING SPROUTABLE MATERIALS

While there are a wide variety of seeds, beans and grains that will sprout successfully, finding the right type for sprouting can be a problem. It is important that the seeds have a high rate of viability. This means that a high percentage of the seeds will actually germinate or sprout. Those that don't sprout either remain hard, tooth-cracking lumps or turn into a rotting mush. Seeds purchased for sprouting should have a minimum specified viability. A good grade of mung beans, for instance, should have a minimum germination rate of 90 percent.

"Seed-quality" beans are generally recommended for sprouters, as compared to "food-quality." Seed quality means that the seeds are meant to be grown and therefore will sprout. Food quality means the seeds were meant for cooking in their dry, unsprouted state, are of a lesser grade, and have a lower germination rate.

Seed-quality products which are designed for farmers often have been treated with chemicals to retard spoilage and combat pests. If you use seeds from farm-supply houses, special care must be taken to wash them carefully.

Food-quality seeds are generally untreated by toxic chemicals but they may have been sprayed with preservatives to increase shelf life. This sometimes hinders their viability and also requires several rinsings before soaking.

Luckily it is becoming easier to purchase seeds, beans and grains specifically grown for sprouting. These can be found in health-food stores and specialty shops, and are available from many excellent mail-order houses.

Since mung beans have little use except for sprouting, they are almost always a reliable choice. Some supermarket beans sprout well, and novices should have a reasonable degree of success with commercial green lentils, black-eyed peas, green peas (not split peas), and chick-peas. It is interesting to try to sprout any bean or seed that you might have on hand or purchased for another use. Try one or two tablespoons to find out how many germinate and if you like the taste.

The prices of sprouting materials vary widely. The following are current prices (price per pound for 1-pound sack) from several of the major mail-order suppliers:

	Mung	Wheat	Soy	Alfalfa
Supplier A	$1.20	$ .35	$ .60	$1.85
Supplier B	$ .54	$ .16	$ .18	$ .89
Supplier C	$1.10	$ .32	$ .39	$1.10

Enthusiastic sprouters buy their beans in quantity and save considerably. But before you get carried away and buy a 100-pound sack of lentils, test a small quantity of the same batch for taste and viability. A list of trustworthy mail-order houses is supplied at the end of this book, but while a mail order house may have very reasonable prices, shipping costs can easily negate any savings. Your best source of materials will probably be a local health-food store that you can trust.

CHOOSING SPROUTS

Every living plant was once a sprout of one sort or another. Experts contend that virtually 99 percent of all vegetation is edible in the sprout stage—but you shouldn't try to cultivate either potato or tomato sprouts, which are said to be poisonous if eaten in quantity.

In early spring, walking through budding woods, I sometimes find tender green sprouts buried in the leafy loam and wonder what they might become: oak, maple, shrub or wildflower? Then, in a childish spirit of adventure, I'm apt to taste a fragile, translucent shoot that conceals the beginnings of a life. The flavors have certainly been exotic, with a wildness and strangeness you generally don't encounter in your own kitchen. Secretly I think I've eaten a jack-in-the-pulpit or blueberry bush or walnut tree. What power! To have eaten the whole tree!

It's a cheap thrill, one with not too many risks. I'm really not very daring, for with all my field guides, I have never positively identified any wild mushroom, even the most common edible varieties, to my own satisfaction. I always worry that I may have made a mistake. But one tiny sprout—well, why not?

Foraging for wild sprouts is a survival skill. I've never thought I'd do particularly well on a desert island, but now I comfort myself that if I couldn't catch a fish with a paper clip or find a flint for a fire, I could always grub about and survive on those incredible sprouts.

Don't worry—there are many reliable and delicious sprouts available without getting down on your hands and knees to make a meal from garden flowers or roadside weed sprouts.

The most common beans, seeds and grains for sprouting are alfalfa, lentil, mung, rye, soy (yellow), wheat.

Other popular varieties include almonds, buckwheat, chia, red clover, corn (yellow Texas), fenugreek, flax, garbanzos (chick-peas), millet, black mustard, unhulled oats, garden peas, unhulled pumpkin, black and red radish, rice, safflower, unhulled sesame, squash, sunflower (hulled and unhulled), turnip.

Some people have reported success sprouting anise, asparagus, basil, barley, beets, caraway, carrots, celery, chives, cress, dill, eggplant, fava beans, kale, kidney beans, lettuce, lima beans, marjoram, onion, parsley, peanuts, black-eyed peas, pinto beans, pumpkin, rutabaga, sage, spinach, swiss chard, thyme.

There's no reason not to experiment. Test seeds planned for your vegetable or flower garden. The only problem is finding seeds that will sprout consistently and sprouts that taste good. I have heard that some people sprout almonds, for instance, but I have never been successful in finding the right almond for sprouting. When trying a new or esoteric sprout, begin by sprouting only one tablespoon of dried seeds as a test case.

SPROUT FAMILIES

There are some sprouts which seem to have the same characteristics in sprouting and behave similarly when used in cooking. I've grouped these together into sprout families. They probably have no botanical relationship, but if you have sprouted one member of a group and know its peculiarities, the others will tend to behave in a like way. More important, when you are using a recipe which calls for a specific sprout, you can usually substitute a member of its sprout family, without having to alter the recipe any further. The tastes will vary of course, but the substitutions will not change drastically the proportions of bulk or water content in a recipe. Ground soybeans, for instance, can be substituted for ground peas or garbanzos and vice versa.

THE SMALL SEEDS. These grow rapidly, taste delicious with the first small leaves, and are excellent in sandwiches and salads. They sprout well in combination with others and with members of their own group. They include:
Alfalfa
Clovers
Millet
Mustard
Radishes
Sesame

THE GRAINS. The grains tend to get sweeter as they lengthen, are used in baking and salads, and are sometimes grown into and eaten as grasses. They include:

Barley
Oats
Rye
Wheat

THE TENDER BEANS. These are the most reliable and can be sprouted well in combinations or in large quantities. They are well liked in their raw state. They include:

Green Lentils
Mung Beans

THE TOUGH BEANS. These beans usually must be steamed, boiled or cooked in some manner before eating. They do not sprout well in combination and must be rinsed and culled frequently to prevent spoilage. They include:

Black-eyed Peas
Kidney Beans
Lima Beans
Navy Beans
Pinto Beans

THE SOY FAMILY. These are similar to the tough beans and require similar care during sprouting and cooking. They are the most nutritious of all sprouts and are worth the trouble of learning to sprout them. They include:

Garbanzos (chick-peas)
Garden peas
Soybeans

THE HEAVY HULLS. The hulls of these sprouts must be removed before they are cooked or eaten. They include:

Almonds
Buckwheat
Peanuts
Pumpkin
Squash

THE MUCILAGINOUS SEEDS. The mucilaginous seeds become sticky when they are soaked in water. Follow this procedure when attempting to sprout them:

1. Pour water into a saucer.
2. Sprinkle seeds over the water and let stand eight hours or overnight.
3. In the morning the seeds will have absorbed the moisture and adhered to the bottom of the saucer.
4. Rinse by running water gently into the saucer and carefully pouring it out. Use a sieve if the seeds are loose.

The mucilaginous seeds include:
Chia
Cress
Flax

SPROUTING TOGETHER

Seeds and beans from completely different sprout families can be sprouted simultaneously to produce interesting and tasty combinations which are cooked and eaten together. The rate of growth is similar for several types of sprouts (as listed below) and, in combination, these are usually ready to harvest at the end of the third day. Sprouting together is a unique way to enjoy the different tastes of sprouts without having to sprout in many separate containers and having many leftovers to store and worry about.

You don't need special equipment or techniques, but, since most combinations include some of the small seeds, you will want to use a wire strainer during the rinsing and draining process to prevent any seeds from slipping out and washing away. Before creating combinations, a few samples of each variety of seed or bean should be tested first for viability. If one member of the team refuses to sprout, you may lose the entire crop to rot. Members of the tough bean, heavy hull, and soy families don't usually sprout well in combination.

You may work out and make up your own combinations to your tastes, needs, and pocketbook or purchase ready-made ones. Here are some common mixtures that sprout well in tandem.

1. Mung, lentil, alfalfa (⅓ each)
   Excellent in salads, sautéed, in soups, stirred into rice. The mung provides the green crunch, the lentil gives a pepper spice, and the alfalfa adds a tender sweetness. A very reliable group to work with.

2. Wheat, rye, and flax (2 parts wheat, 2 parts rye, to 1 part flax)
   A delicious breakfast mixture. Serve when all the sprouts are just barely as long as the grain itself.

3. Alfalfa and black radish (1 part alfalfa to 1 part black radish)
   The sweet alfalfa combined with the tart radish makes a delightful combination in salads or used with mayonnaise in sandwiches. Stir into tuna, chicken or seafood salads in place of onion and celery.

4. The Supermix: Alfalfa, black radish, green lentils, mung beans, and red clover (any combination)
   A great way to use little bits of seeds and beans left over from previous sproutings. Use any amounts on hand. This makes a wild combination to toss into salads, soups, Chinese dishes or separately as its own vegetable. As you chew, the flavors change and excite the palate. People won't be able to figure out exactly what they are eating—but they'll ask for more. Expose sprouts to light for a few hours to bring out the multi-shades of green in all the various sprouts.

A DICTIONARY OF BEANS, SEEDS AND GRAINS

ALFALFA (*Medicago sativa*)

Alfalfa brings to mind a grassy barnyard food, but its flavor when sprouted is truly delectable. Alfalfa sprouts are rich, sweet and nutlike.

The Arabs discovered that when they fed this grain to their famous horses, the horses became stronger and swifter. Then they tried it themselves and found it improved their own physical prowess, so they named it *al-fal-fa*, or father of all foods.

Today alfalfa is widely recognized as being the best possible fodder. Its roots penetrate deep into the earth searching for all the nutrients and elements it craves. The alfalfa plant, which grows to a maximum of 2 to 4 feet, has a root system running hundreds of feet underground.

This sweet sprout tastes so delicious that one would think it had to be fattening. It isn't! It has a high mineral value, containing phosphorus, chlorine, silicon, aluminum, calcium, magnesium, sulphur, sodium and potassium. The sprouts are rich in vitamins C and K, and when the tiny primary leaves are exposed to light, they rapidly develop chlorophyll.

Alfalfa seeds generally have a high rate of germination and sprout easily in combination with others. The seeds are among the most expensive, but a few go a long way. When dried, alfalfa sprouts have a nutlike flavor and can be used in place of nuts in baking.

HARVEST: 1″ to 2″, or just after the first leaves appear.

Almonds (*Prunus amygdalus*)

To sprout, you will need fresh, unhulled almonds. They must be soaked at least 12 hours and kept very moist until they sprout in about 4 days. They are used primarily in making almond milk, in which 1 cup of sprouted almonds is blended with 1 quart water. Chopped, they may be added to ice cream, cakes, cookies, and salads.

HARVEST: ½" to 1".

Barley (*Hordeum vulgare*)

This hardy cereal grass has been cultivated since prehistoric times. It is important to find the unhulled barley for use in sprouting.

HARVEST: When shoot is about the same size as the grain.

Beans (*Phaseolus*)

Most members of the bean family will sprout if viable beans can be found. Try beans from the supermarket, but be certain to remove beans that don't show evidence of sproutability. Most sprouted beans need steaming before they are tasty or digestible. Try grinding the sprouted beans and adding them to meat dishes, eggs and sandwich spreads. Beans that sprout well include navy, jack, kidney, pinto, fava, and lima.

HARVEST: Large beans must be harvested very young, while the sprouts are still immature—at about 2 days old or ¼" to ½" long.

Buckwheat (*Fagopyrum esculentum*)

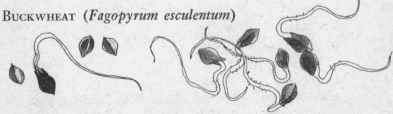

Buckwheat has been traditionally used as fodder, and also as flour by the most devoted pancake lovers. The

buckwheat sprout is one of the few that must be hulled before it is eaten. It is rich in vitamin B complex and vitamin E, and contains large amounts of rutin. Follow the directions for growing wheat.
HARVEST: 1″ to 1½″.

CHIA (*Salvia columbariae*)

This mucilaginous seed is a Mexican herb of the mint family. It has long been a staple of the Indians, who used chia to sustain them on long marches and migrations. It is mild-tasting, similar to flax, and is excellent in cereal combinations. Try stirring the sprouts into any hot or cold cereal or sprinkle them on fruit. Follow directions for mucilaginous seeds (page 42).
HARVEST: ¾″ to 1″.

CHICK-PEAS (*Cicer arietinum*)
See Garbanzos

CLOVER, RED (*Trifolium pratense*)

This common forage plant makes a delicious salad and sandwich sprout, similar to alfalfa. It is an excellent source of chlorophyll after the first primary leaves have developed and have been placed in sunlight for a few hours.
HARVEST: ¾″ to 1″.

CORN (*Zea mays*)

The variety of sweet corn that sprouts best is Texas Deaf Smith County.
HARVEST: ¼″ to ½″.

CRESS (*Lepidium sativum*)

Cress is an herb of the mustard family. Its sprouts are tasty in salads and sandwiches. Follow the sprouting directions for mucilaginous seeds (page 42).
HARVEST: ¾″ to 1″.

Fenugreek (*Trigonella foenum-graecum*)

Fenugreek, meaning Greek hay, is a strong-scented Old World herb of the pea family. It is known for its medicinal qualities and as a component of curry powder. The strong soak water makes an interesting herb tea. Use the fenugreek sprouts in salads, soups, curries, sandwiches and rice combinations.

HARVEST: Just sprouted, ¼″. Do not let sprouts get too long, or the taste will be extremely bitter.

Flax (*Linum usitatissimum*)

Flax, or linseed, is known for its bark, used to make linen, and its oil, used medicinally for soothing and softening. Flax sprouts were used by ancient Romans and Greeks as a delicacy served between banquet courses. Flax is excellent sprinkled on breakfast cereal. Follow sprouting directions for mucilaginous seeds (page 42).

HARVEST: ¾″ to 1″.

Garbanzos (*Cicer arietinum*)

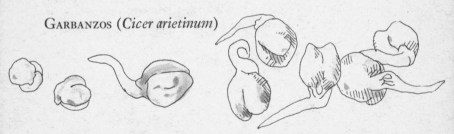

The garbanzo is also known as the chick-pea in the United States, ceci in Italy, and Bengal gram in India. It is commonly used in Asia and Latin America as a vegetable, the seeds being eaten boiled or roasted like peanuts. When garbanzos are very thoroughly roasted, they make an acceptable substitute for coffee. Garbanzos are very high in protein and can replace soybeans in recipes.

Steam the sprouts before using in salads, sandwiches or other dishes that will not be heated.

HARVEST: ½" to ¾".

LENTILS (*Lens esculenta*)

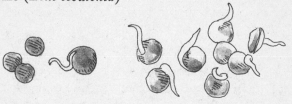

The Bible mentions that Esau sold his birthright to Jacob for a mess of red pottage made from lentils; and they were a staple of the early Egyptian diet. Lentils are very inexpensive and very nutritious, as they contain iron, cellulose and the B vitamins. Lentils make a very reliable sprout and most ordinary supermarket lentils will sprout successfully if you remove dark or broken seeds. The flavor of the raw sprouts is like that of fresh-ground pepper on salad greens; cooked, their flavor is more nut-like. Lentils sprout well in combination with other seeds and in large quantities. They are perfect in salads, soups, vegetable combinations, Chinese cooking and as a general substitute for celery or green pepper.

HARVEST: When the shoot is no longer than the seed.

MILLET (*Panicum miliaceum*)

Millet is another forage plant that makes a good sprout. It is high in protein, calcium, riboflavin and essential amino acids. Millet is one of the easiest grains to digest and acts as an intestinal lubricant. As a sprout it makes an excellent cereal base; it is also used in baking.

HARVEST: When the shoot is the same size as the grain.

Mung Beans (*Phaseolus aureus*)

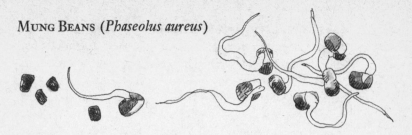

The mung bean, also called the gram (green, black, and golden), is the most important bean for sprouting. The first record of the plant is contained in the medical writings of the Emperor Sheng-Nung in 2383 B.C. The mung bean doesn't seem to be cultivated for any use except sprouting, and almost any mung bean can be relied upon to produce a good sprout. For this reason, they are a good first choice for a beginner to sprout.

Mung sprouts are rich in calcium, phosphorus, iron, vitamins A and C and are a staple of many oriental diets. In Chinese cooking, the hulls are usually removed but this tedious task can be eliminated, as the hulls are easily digestible and are rich in minerals. They may be harvested at almost any time from the moment the shoot appears until the sprouts are 4 inches long and are juicy and plump. But they will dry out and shrivel if left to grow longer than that.

Mung beans can be used in many dishes—salads, vegetables, main dishes and Oriental cooking. They are delicious raw, but a few seconds' steaming will remove any "green" taste. In cooking, add them only at the very last moment, so they heat through but do not wilt.

HARVEST: ½ " to 4".

Mustard (*Brassica*)

The black mustard is the most common variety used for sprouting. It has a tangy, spicy taste, not unlike fresh English mustard. The mustard seeds of sprouting quality are among the most expensive, but a few go a long way in adding a delectable flavor to salads and sandwiches.

HARVEST: ½ " to 1".

49

OATS, UNHULLED (*Avena sativa*)

This common forage plant and cereal grass makes a delicious sprout. Use in baking, on cereal, in salads and vegetable combinations.

HARVEST: When the shoot is no longer than the grain.

PEAS (*Pisum sativum*)

The sprout of this climbing legume can be used in recipes interchangeably with the soybean sprout. The sprouts are best when lightly steamed before they are added to salads and other raw dishes. They are delicious ground or puréed or simply boiled and served with butter.

HARVEST: ¼″ to ½″.

RADISH (*Raphanus sativus*)

Both the red and black radish, members of the mustard family, make good sprouts. They have a tangy, peppery taste, like a mild sliced radish, and are delightful in salads and sandwiches. They are easy to sprout and work well in combinations. Although they are among the most expensive seeds for sprouting, only a few need to be added to a dish. Claims have been made that the black radish sprout has aphrodisiac qualities, although I haven't been able to find proof of this—yet!

HARVEST: ½″ to 1″.

RYE (*Secale cereale*)

This cereal grass is closely allied to wheat. The rye sprout is used frequently in breads, cookies, desserts, as a cereal, with fruit and in salads. Rye is rich in vitamin E, phosphorus, magnesium and silicon. The rye sprout becomes sweeter as it lengthens.

HARVEST: For eating raw: when the shoot is the size of the grain. For sweeter sprouts and for cooking: up to 1", or let grow into rye grass.

SESAME OR BENNE (*Sesamum indicum*)

Sesame is a popular plant because of its almondlike flavor. It is nutritionally high in protein, unsaturated oil, calcium, phosphorus, magnesium, niacin and vitamins A and E. The sprouts are used as a breakfast food and are often blended into beverages. They have a sweet almond flavor when eaten young, but become bitter if the sprouts grow longer than 1/16". Unhulled sesame seeds must be used for sprouting.

HARVEST: When sprouts just show the barest bud. Usually ready in 2 days.

SOYBEANS (*Glycine max*)

The soybean is undoubtedly the most nutritious of all the beans for sprouting, and it has been the major source of protein in Oriental diets for centuries. The yield of protein, weight for weight, is approximately twice that of meat, 4 times that of eggs, twice that of lima beans and most nuts, and 12 times that of milk. The soybean protein is considered superior in quality to that found in other legumes because it contains the essential amino acids found in proteins of animal origin.

Soybean sprouts release significant quantities of vitamin C not found in the dried bean. They are also rich in vitamins A, E, and B complex, and contain calcium, phosphorus and iron, as well as lecithin, which plays an important role in the absorption and transport of fats in the bloodstream.

The soybean is not the easiest bean to sprout, especially in hot weather. The yellow variety is best. Look for Chief, Ebony, Illini, Lincoln and Richland, as well as those varieties labeled specifically for sprouting. Soybeans require frequent rinsings; I recommend a minimum of 3 times daily.

Some cooks insist on removing the slippery outer husk, and the raw bean flavor may be too strong for most tastes. Try steaming or boiling the sprouts for about 5 minutes; they will acquire a nutlike taste. Whole, they are excellent additions to salads, casseroles and egg dishes. Ground, they may be added to meat loaves, stuffings, breads, as a base for sandwich fillings, dips, salad dressings and vegetarian dishes. Soy sprouts make a good substitute for those crisp Chinese vegetables that are difficult to locate in supermarkets. They do not sprout well in combination with other seeds.

HARVEST: 1/4" to 1 1/2".

SUNFLOWER (*Helianthus*)

Sunflowers were first used as a food by American Indians, who taught the early settlers how to grow them. The nutritional powerhouse in the seeds is due in part to the flower's ability to follow and face the sun from morning to night, absorbing the maximum amount of life-giving sunshine. The plant's extensive root system extracts many trace minerals not found in topsoil. The seed contains phosphorus, calcium, iron, fluorine, iodine, potassium, magnesium, zinc, vitamins D and E, unsaturated fatty acids and about 30 percent protein.

Sunflower seeds should be sprouted in a warm place (75°–85°), and they take longer than average to mature, usually 5 to 7 days. Sunflower sprouts must be eaten when very tiny or they will be bitter. The husk is always discarded before they are eaten. You can also grow hulled sunflower seeds specifically recommended for sprouting.

Pumpkin and squash seeds are sprouted in the same manner.

HARVEST: When shoot is just showing, barely budded. Not over ½".

WHEAT (*Triticum sativum*)

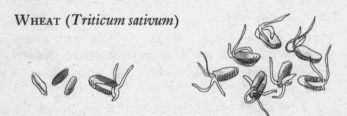

This most important cereal grass is an excellent sprouter. It contains protein, vitamin E, and large amounts of nitrates in the husk. It is used whole in baking and in salads, and added, ground, to many other recipes. As wheat lengthens, it sweetens and is used then in desserts and with fruits. Sprouts are grown and harvested the traditional way.

HARVEST: Same as for Rye, page 51.

General Cooking Directions

POST-HARVEST PREPARATIONS

You have hovered over your sprout garden for almost a week. You have faithfully soaked, drained and harvested the sprouts. Now, what in the world do you do with those funny squiggly vegetables?

Before eating sprouts raw or cooking them, a final rinsing, cleaning and culling is in order. Place the sprouts in your strainer a handful at a time and rinse with a light spray of water to avoid breaking the tender shoots. Drain completely to avoid adding any excess water to your recipes.

Don't be afraid of the whiskers that appear on the sprout rootlet. They are not mold, but rather the feeder roots looking for nourishment. These are most apparent in some of the grains, such as rye or wheat, whose roots tangle together in a knotted mass after the third day. Separate them gently after rinsing them.

It is important to check over the sprouts once more, carefully removing those that show no signs of having sprouted. These are often found at the bottom of the container and are generally hard as a stone and not something you would want to chew. Also remove any mushy or broken beans. Now is the time to hull the sprouts if you decide to spend the time and effort, if you find them hard to digest, or if you are using a heavy-hulled seed.

MEASURING

Measure out the amount of sprouts you wish to use in a recipe and store the remainder in the refrigerator immediately. Sprouts are measured by placing them in a measuring cup and lightly pressing until the cup is filled to the desired amount. Don't push so hard that they crush or break, but don't fill the cup too loosely, leaving too much air space between the sprouts. If you have measured properly, the sprouts will pop back up after they are pressed in for a moment. A half-cup of mung beans, loosely packed, will measure ¾ cup, but they can easily be pressed to the ½-cup mark without bruising the sprouts.

ADJUSTING RECIPES

When a recipe calls for one particular type of sprout, you may experiment by substituting a sprout from the same family group which will behave in a similar manner in a cooked recipe. In salads, soups and sandwiches it rarely matters what substitution is made. But since no two sprouts taste alike, the flavor will differ in the finished recipe.

Remember that sprouts contain large amounts of water which must be taken into account when experimenting and adding sprouts to your own recipes. This is especially important in baking breads.

When grinding raw sprouts, choose immature ones (usually 2 days old) which haven't absorbed too much water yet. But when using sprouts raw in salads and sandwiches, the older, plumper sprouts add more crispness and flavor.

STORAGE AND PRESERVATION

The raw seeds have a very long shelf life. If kept free from bugs and rodents and away from moisture, they can truly last indefinitely. Seeds found in sealed Egyptian tombs have sprouted to life after 5,000 years of dormancy. Since the seeds and beans are usually sold in plastic or paper wrappings, I suggest you transfer them to glass or heavy plastic storage containers before putting them away in the cupboard. Label as to variety and date purchased. If you are testing many kinds, you will need have on hand only small amounts of each. When you discover favorites, it will pay to order large quantities and save considerably.

The sprouted bean has a refrigerator life of 7 to 10 days, depending on the variety. For the first 7 days the sprouts show a steady increase in the amount of vitamin C, *even after refrigeration*. From that time on, they begin to lose their potency.

Refrigerate the sprouts immediately when they reach their peak for harvesting. First rinse them quickly in cold water, drain thoroughly and wrap them loosely in a single layer of damp paper toweling. Place the sprouts in a plastic bag and seal tightly. If you aren't using them up quickly and they begin to wilt or dry out, they may be rinsed again in cold water, rewrapped and refrigerated.

Since sprouts are so inexpensive and can be grown from scratch so rapidly, it doesn't pay to keep old sprouts or attempt to freeze, can or preserve them. If you can't bear to throw anything out and are unable to use them up, sprouts can be placed in plastic freezer boxes and frozen for a few months.

Sprouts may be dried very successfully. Spread them on a cookie sheet and leave in a warm room or place in a slightly heated oven until they are dry. Grind the dried sprouts in a blender and store in a tightly covered jar. This nutty, delicious sprout powder can be used as an additive to beverages, baked goods, baby foods, desserts, nut butters, spreads, etc. It is a nutritious food concentrate, with more food value than the original dry seed, and will keep for a long time. Wheat, rye, soy, sesame and alfalfa are all excellent candidates for drying.

Recipes

APPETIZERS

Sprout Fritters

These fritters make a wonderful appetizer, vegetarian main dish or vegetable. Ask your guests to figure out the ingredients that create the interesting flavors and textures.

- 2 cups sprouted wheat, rye or alfalfa
- 1 cup almonds, walnuts or pecans
- 1 large onion
- 2 cups whole-wheat bread crumbs
- 1 tablespoon salt
- 2 tablespoons vegetable oil
- 1 cup milk
  Oil for deep-fat frying
  Black pepper

1. Put the sprouts, nuts and onion through the small blade of a meat grinder.
2. Stir in bread crumbs, salt, oil and milk.
3. Shape into balls about the size of a walnut.
4. Heat oil for frying. The pan should contain at least 1½ inches of hot oil.
5. Fry fritters in small batches in very hot oil. Turn them gently to brown evenly. Remove with a slotted spoon when brown and crisp—about 30 seconds to 1 minute.
6. Drain on paper towels. Grind some fresh pepper over fritters before serving.
7. Serve with parsley topping, in a cream sauce or meat gravy.

YIELD: About 30 fritters.

Quick Cream Cheese Puffs

    6 frozen patty shells, baked according to directions.
    8 ounces cream cheese, at room temperature
    2 tablespoons softened butter
    1 egg, beaten well
    3 drops Tabasco sauce
    2 tablespoons chives
    ¼ cup sprouts (alfalfa, radish, mustard)

1. Place shells on baking tray.
2. Mix cheese, butter, egg, Tabasco, chives and sprouts together.

3. Fill each patty shell and bake in 400° F. oven for about 25 minutes, until cheese is puffed up and slightly brown on top.

YIELD: 6 servings.

Sprout Butter I

1 cup sprouts, chopped or ground
1 cup butter, softened
   Herbs and seasonings (dry basil, tarragon, curry powder, soy sauce, etc.)

Combine all ingredients and serve as basis for hors d'oeuvres.

Sprout Butter II

Toasted sprouts (soy, garbanzo, pea)
Vegetable oil
Salt to taste

Grind sprouts in blender, adding oil drop by drop until mixture is the consistency of peanut butter. Use like peanut butter.

Cheese Snacks

½ cup sprouts (alfalfa, clover, mustard or radish)
½ cup cream cheese
   Dash curry powder
½ teaspoon salt (if nuts or seeds are unsalted)
½ cup chopped nuts, or toasted sunflower seeds

1. Mash sprouts, cheese, curry powder and salt together.
2. Shape into balls.
3. Roll into nuts or seeds.
4. Chill and serve.

YIELD: A snack for 3 or 4.

Sprout Cracker Spread

This makes a good hors d'oeuvre base when spread on crackers and topped with chopped eggs, olives, meats or seafood; with a little more oil stirred in, it can be used as a dip.

  1 cup alfalfa, clover or black radish sprouts
¼ cup garbanzo sprouts
½ cup mung bean sprouts
  2 to 3 tablespoons oil
  1 small onion
  1 tablespoon lemon juice
  1 teaspoon salt

Run sprouts through grinder. Blend in oil, onion, lemon juice and salt.

YIELD: 1 cup spread.

BEVERAGES

Soy Sprout Milk

Unbelievably nutritious, this milk can be used as a substitute for cow's milk in many recipes; it makes a delicious drink.

1 cup sprouted soybeans
4 cups warm water
2 tablespoons honey

1. Blend beans and water in blender with honey for 5 minutes.
2. Cook over medium heat about 10 minutes, stirring constantly.
3. Strain. Let cool and use strained liquid as beverage, sauce base, soup stock, etc. Use the sieved residue as a filler in meat and vegetable dishes.

Sesame or Sunflower Milk

1 cup sprouted sesame or sunflower seeds (shoot has just budded)
2 cups cold water
2 tablespoons honey

1. Blend together in blender for 4 minutes.
2. Put through sieve.
3. Serve plain over ice, or add rum, coconut, pineapple juice or almond extract for added flavor.

Sweet Sprout Cocktail

1 cup alfalfa sprouts
2 cups unsweetened pineapple juice
1 tablespoon honey

1. Blend together in blender for 3 minutes.
2. Strain, if desired. Serve with ice.

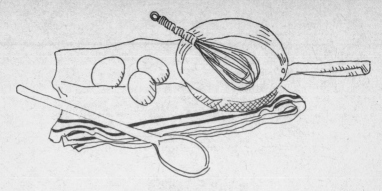

BREAKFASTS

Sprout Omelet

    1 cup sprouts (mung, wheat, rye, alfalfa)
    2 to 3 tablespoons butter
    4 eggs, beaten
    2 tablespoons water
  ¼ teaspoon salt
      Pepper

1. Sauté sprouts in butter for 2 minutes. Remove.
2. Blend eggs, water, salt, and pepper in a bowl.
3. Clean out skillet with paper towel. Heat.
4. Add more butter to pan if necessary. Pour eggs into skillet, cook slowly, running spatula around edge to allow uncooked portion to flow underneath.
5. Sprinkle sprouts on top of cooked eggs. Fold over and turn onto platter.

YIELD: 2 to 3 servings.

Wake Up with Sprouts

    1 cup sprouts (wheat, rye, alfalfa)
    1 cup sunflower seeds, hulled
    3 lemons, juiced
    3 tablespoons nutritional yeast
    3 tablespoons honey

½ cup powdered milk
6 apples, grated
½ cup wheat germ
¼ cup coconut (optional)

Mix all together and serve.

YIELD: 4 to 6 servings.

Mock Grape-Nuts Cereal

1 cup dried alfalfa sprouts, ground in blender
¼ cup raisins

Mix raisins with sprouts. Serve with sugar to taste and milk or cream.

YIELD: 1 serving.

Sprouts and Oatmeal

Add 1 cup sprouts (dried wheat, rye, or alfalfa) to 1 cup oatmeal and cook according to directions for the oatmeal.

YIELD: 6 servings.

Breakfast Sprout Salad

2 cups wheat or rye sprouts (3 days old)
1 cup hulled sunflower seeds
3 apples, grated
3 bananas, sliced thin
½ cup currants or raisins
½ cup yogurt

Blend all ingredients together. Sweeten with honey, if necessary.

YIELD: 4 to 6 servings.

BREADS

The Russians discovered that if they soaked their wheat overnight before grinding, their bread was so much more nutritious than bread made with regular milled dry wheat, that eating it would help combat scurvy. But using sprouts in baking presents certain problems. The moisture in the sprouts may cause an ordinary recipe to flop.

Care must be taken to use sprouts that are young, without too much growth in the shoot. Sprouts must be carefully drained, and preferably patted dry with toweling before grinding.

To your favorite bread recipe add ½ to 1 cup of sprouts in place of an equal amount of flour and liquid. 1 cup wheat sprouts would displace ½ cup of flour and ½ cup of liquid.

You can usually toss in a few young whole sprouts without adjusting a recipe; just knead them in at the last moment. But pick off the sprouts that come to the top surface of the bread before putting it in the oven, or they will scorch, adding a burnt taste to the crust.

I have found that breads containing sprouts take at least 10 percent longer to bake. Expect some trial and error for the first few loaves, but the end result will be a much appreciated healthful fresh loaf of bread.

Sprouted Sandwich Loaf

The crunchy, nutty sprouts provide an interesting contrast in this fine-textured white bread.

    1 cup wheat, rye or alfalfa sprouts
    1 cup milk
    1 tablespoon butter
 1½ tablespoons shortening
    2 tablespoons sugar
    1 cup cold water
 2½ teaspoons salt
    1 package yeast dissolved in ¼ cup water at 85° F.
 6½ cups unbleached white flour

1. Rinse and drain sprouts. Lay them on a piece of paper toweling in a thin layer. Cover with a second layer of towel. Roll gently into cylinder shape and set aside. This will remove all excess moisture from the sprouts.
2. Scald the milk. Remove from heat and add butter, shortening, and sugar. Stir till fat melts.
3. Add water and salt. Let cool until temperature has reached 85° F.
4. Add yeast dissolved in water.
5. Slowly stir in sifted flour. When all ingredients have been combined, let dough rest 10 minutes.
6. Knead until smooth and elastic on lightly floured surface. Place in greased bowl, turn to grease entire surface of dough. Let rise in warm place (75° to 85° F.) until double in bulk.
7. Punch dough down on floured surface. Flatten in large circle. Sprinkle sprouts on dough. Remove any seeds that haven't sprouted. Knead sprouts into dough. Keep turning dough and adding sprouts, until they are evenly distributed.
8. Separate dough into 2 portions. Form into loaves.
9. Place in greased loaf pans and let rise again in warm place until almost double.

10. Bake in oven that has been preheated to 400° F. for 10 minutes. Reduce heat to 350° F. and bake 20 to 30 minutes longer.
11. Remove loaves at once from pans. Cool on rack.

YIELD: Two 5″ x 9″ loaves.

Sprouted Wheat Bread

Use sprouts that are young; the shoot must not be longer than the grain itself.

    3 cups lukewarm water
    2 tablespoons dry yeast
    1 tablespoon salt
    ¼ cup honey
    3 tablespoons vegetable oil
    3½ cups unbleached white flour
    1 cup ground wheat sprouts
    1 cup wheat sprouts, whole
    2 cups whole wheat flour

1. Pour 1 cup lukewarm water into large bowl. Add 2 tablespoons yeast and dissolve.
2. Add the remaining 2 cups of water, the salt, honey and oil.
3. Stir in white flour. Beat well. Cover and let sponge double in warm place (80° F.).
4. Add ground and whole sprouts to sponge. Work in about 2 cups whole wheat flour. Knead until smooth and elastic. Place in clean oiled bowl, cover, and let rise again in warm place until doubled.
5. Knead again, adding more flour if necessary. Form into 2 loaves and place in greased pans. Bake at 350° F. 1 to 1¼ hours.
6. Remove from pans. Cool on wire rack.

YIELD: 2 loaves.

Alfalfa Bread

It is important to use young sprouts; the higher water content of older sprouts will result in a soggy bread.

      2 small potatoes with skins
      1 quart water
    ½ cup honey, plus 2 tablespoons
   1½ tablespoons salt
    ¼ cup vegetable oil
      2 packages dry yeast
   8 to 10 cups flour, whole wheat, white or combination of these
      4 cups 3-day alfalfa sprouts

1. Wash potatoes well, cut into small pieces and cook in 1 cup water until done. Liquefy in blender with 2 cups cold water. Put through strainer to remove skin particles. Measure. Add enough warm water to make a total of 3½ cups liquid.
2. Add ½ cup honey, salt and oil.
3. Dissolve yeast and 2 tablespoons honey in ½ cup lukewarm water. Let stand 10 minutes and add to potato liquid.
4. Stir in 5 cups flour. Add sprouts. Add additional flour to make a stiff dough. Knead well until smooth and elastic. Dough must not be sticky.
5. Cover with hand towel. Let rise in warm place (80° F.) until double.
6. Punch down and form into 2 loaves ¾ the height of the pans. Place in greased pans.
7. Bake at 350° F. for 1 to 1½ hours, until well done.

YIELD: 2 loaves.

Sprout Bran Muffins

    2 cups whole wheat flour
1½ cups bran
    2 tablespoons sugar
  ¼ teaspoon salt
1¼ teaspoons baking soda
    2 cups yogurt
    1 beaten egg
    4 tablespoons melted butter
    1 cup alfalfa sprouts

1. Combine flour, bran, sugar, salt and soda.
2. Beat together yogurt, egg, and butter and add to dry ingredients. Mix well.
3. Stir in sprouts.
4. Fill well-greased muffin tins. Bake in 350° F. oven about 25 minutes until firm.

YIELD: 20 to 24 muffins.

Sprout Refrigerator Rolls

     1 package dry yeast
3½ to 4 cups white flour
     1 teaspoon thyme
     2 teaspoons celery seed
1¼ cups milk
  ¼ cup shortening
  ¼ cup sugar
     1 teaspoon salt
     1 egg
     1 cup sprouts (wheat, rye, alfalfa)

1. Combine yeast, 1½ cups flour, thyme and celery seed.
2. Heat milk, shortening, sugar and salt together, stirring until the shortening is melted.
3. Add milk mixture to dry ingredients. Add egg and

sprouts. Beat 3 minutes at high speed, or beat by hand until very well blended.

4. By hand, add enough additional flour to make soft dough.
5. Place dough in greased bowl, turning once to grease surface. Cover and chill a minimum of 3 hours.
6. About 2 hours before serving, shape into 3 1" balls (for each roll) and place in cloverleaf fashion in greased muffin tins. Repeat until all dough is used up.
7. Bake at 400° F. for 12 to 15 minutes until light brown.

YIELD: 18 to 24 rolls.

MAIN DISHES

Beef and Bean Sprouts

Mung beans are usually used in this typical Chinese dish, but why not experiment and add green lentils or soybeans or wheat sprouts?

CORNSTARCH MIXTURE:
    1 teaspoon sugar
    1 tablespoon cornstarch
    ½ cup chicken or beef broth
    2 tablespoons soy sauce

Stir the above ingredients and set aside.

    ¼ cup vegetable or peanut oil
    ⅛ teaspoon salt
    1 clove pressed garlic
    1 thin slice fresh ginger, finely chopped or grated
    ¾ pound lean beef, sliced thin while partially frozen
    1 to 2 cups bean sprouts
    4 scallions, chopped, including green ends
    ⅛ cup chopped sweet pickles
    2 tablespoons sherry
    ¼ teaspoon ground white pepper

69

1. Heat a wok or skillet until very hot. Add ⅛ cup oil.
2. Add salt, garlic and ginger with the beef. Stir-fry until just browned for 1 minute.
3. Remove beef and set aside.
4. Add another ⅛ cup oil to hot pan. Add sprouts, scallions, pickles. Stir-fry for 1 minute.
5. Add sherry. Cover and cook 1 minute more.
6. Stir in the browned beef and the cornstarch mixture. Add white pepper. Remove from heat. Stir all together and serve with rice.

YIELD: 3 to 4 servings.

Beef with Radishes and Sprouts à la Staples

⅓ cup cider vinegar
4 tablespoons water
½ cup brown sugar
1 tablespoon cornstarch
1 pound sirloin tip, sliced thin
1 teaspoon cornstarch
2 tablespoons soy sauce
2 tablespoons vegetable oil
1½ cups radishes, sliced thin crosswise
¼ cup scallions cut into ¼″ pieces
1 cup sprouts (mung, lentil)

1. Make sweet-sour sauce by stirring vinegar, water, sugar and the tablespoon of cornstarch together. Set aside.
2. Combine the 1 teaspoon of cornstarch with the soy sauce. Toss the beef slices in the mixture until all are coated.
3. Heat oil in skillet or wok. Sauté meat 1 minute and remove from skillet.
4. Add sauce and radishes to skillet and cook 3 minutes.
5. Add scallions and sprouts. Cook, stirring 1 minute.
6. Add meat and heat another minute or two. Serve.

YIELD: 2 to 3 servings.

Meat Loaf with Sprouts

This makes an extra-nutritious meat loaf. The texture is light, the meat stays very juicy. An excellent cold dish, it slices well for sandwiches.

1½ pounds beef, pork, and veal mixture, ground twice
1½ cups garbanzo or soy sprouts (packed down to measure)
2 beaten eggs
2 tablespoons parsley
1 medium onion
1 teaspoon salt
¼ teaspoon pepper
1 teaspoon lemon juice
½ cup bread crumbs
⅓ cup wheat germ

1. Put sprouts and onion through grinder.
2. Mix with ground meat and all other ingredients.
3. Form into loaf and place in lightly greased loaf pan.
4. Bake at 350° F. about 1 hour.
5. Serve with tomato or cheese sauce.

YIELD: 4 to 6 servings.

Chicken Breasts Sweet and Hot

The chicken is so tender, and the combination of flavors so unusual in this Chinese recipe.

    3 chicken-breast halves, skinned and boned
    2 teaspoons cornstarch
      Half of 1 beaten egg white—beat lightly till frothy, then divide
    3 tablespoons sherry
    ¼ teaspoon salt
    ⅓ cup oil
    3 cups alfalfa sprouts (or combination mung and alfalfa)
    2 teaspoons sugar
      Dash cayenne pepper

1. Slice chicken in strips as thin as possible.
2. Blend cornstarch, egg white, 2 tablespoons sherry, and salt. Toss strips in mixture. Refrigerate at least ½ hour.
3. Heat oil and add chicken. Stir-fry about 1 minute until chicken cooks a little.
4. Remove from pan and drain chicken in sieve.
5. Add sprouts to hot oil and stir-fry 1 minute.
6. Add cooked chicken to sprouts. Add sugar, pepper and 1 tablespoon sherry. Heat through and serve.

YIELD: 2 to 3 servings.

Main Dish Fried Rice

5 tablespoons vegetable oil
2 eggs, beaten lightly
1 cup diced roast pork
1 cup diced cooked shrimp
1 cup diced cooked chicken
2 cups sprouts (mung, soy, lentil)
2 scallions, chopped, with green ends
½ teaspoon pepper
1 cup soy sauce
4 cups hot, cooked rice
1 cup chopped lettuce

1. Scramble eggs in 2 tablespoons oil over medium heat until the eggs begin to set. Remove from pan and set aside.
2. Add 3 tablespoons oil to pan and heat. Add pork, shrimp, chicken, sprouts, scallions, pepper and soy sauce. Stir-fry 2 minutes.
3. Add rice and stir well.
4. Add lettuce and eggs. Stir to heat.

YIELD: 4 large servings.

Egg Foo Yung

There are many variations on this dish. Use any sprout that strikes your fancy and add any leftovers on hand, such as bits of chicken, shrimp or pork. The basic recipe serves 2, but is easily doubled or tripled.

1 tablespoon chopped green pepper or celery
1 small onion, chopped
3 eggs
½ to ¾ cup sprouts (precooked if soybeans are used) or combine sprouts with leftovers to equal ¾ cup total volume.
½ teaspoon salt
¼ teaspoon ground ginger
Oil or margarine for frying

1. Brown pepper or celery and onion till soft.
2. Beat eggs until they are frothy.
3. Add pepper, onion, sprouts and leftovers. Stir in salt and ginger. Mix well.
4. Heat 1 tablespoon fat in 6″ skillet. Pour ½ cup of egg mixture into skillet and brown quickly on both sides. Repeat until egg mixture is used up.
5. Serve with rice and soy sauce or Egg Foo Yung Gravy (below).

YIELD: 2 to 3 omelets.

Egg Foo Yung Gravy

1½ cups chicken stock
1 tablespoon cornstarch
2 tablespoons soy sauce
⅛ teaspoon pepper
Pinch of salt

Mix all ingredients together in saucepan. Heat slowly, stirring constantly, until thick. Keep warm until ready to use.

YIELD: Enough gravy for 6 omelets.

Shrimp and Vegetables

2 tablespoons soy sauce
½ teaspoon salt
1 teaspoon ground ginger
1 tablespoon sherry
1 pound uncooked shrimp, shelled and cleaned
3 tablespoons vegetable oil
1 cup chopped onion
¾ cup chopped celery
1 can (6 ounces) water chestnuts, sliced
3 cups bean sprouts (mung or lentil)

1 tablespoon cornstarch
1 cup chicken broth

1. Mix soy sauce, salt, ginger and sherry together and sprinkle on shrimp.
2. Heat 1 tablespoon oil in skillet, add shrimp and sauté about 7 minutes. Set shrimp aside.
3. Heat the remaining 2 tablespoons oil in skillet, add onion and celery, and sauté about 5 minutes.
4. Add shrimp, water chestnuts and sprouts.
5. Mix cornstarch and broth together. Stir into rest of ingredients and cook, continuing to stir, until thickened.
6. Cover and simmer about 3 more minutes.

YIELD: 6 to 8 servings.

Sprouts, Pork and Peppers

Watch out! This dish is *hot*. Try this along with a few other Oriental dishes served at the same time, if you have any sensitive palates to please.

¼ cup (or more) vegetable oil
1 pound pork, cut into very thin strips
¼ cup seeded and shredded hot peppers
2 tablespoons soy sauce
3 cups sprouts (mung, lentil, soy)
½ teaspoon salt
1 teaspoon sugar
1 tablespoon sherry

1. Heat ¼ cup oil in skillet. Add pork and peppers. Stir-fry for 2 minutes, or until pork strips are completely white. Add soy sauce. Cook 30 seconds.
2. Add more oil to pan if sticking.
3. Add sprouts. Cook 1 minute.
4. Stir in salt, sugar and sherry. Serve hot.

YIELD: 2 to 3 main-course servings.

Crunchy Chicken Legs

    1 cup ground, roasted (see page 88) sprouts (soy,
        pea, garbanzo)
    1 beaten egg
    ¼ cup milk
    1 teaspoon salt
    6 chicken legs

1. Spread ground sprouts on platter.
2. Combine egg, milk and salt.
3. Dip each leg into egg mixture, then roll in sprouts.
4. Lay on greased baking sheet and bake at 350° F. about
   45 minutes, or until chicken is tender.
5. Serve warm or cold as picnic dish.

   YIELD: 2 to 3 servings.

Carrie's Rock Cornish Hen Stuffing

    1 small potato
    1 stalk celery
    1 tablespoon soy sauce
    ½ cup bean sprouts
    1 tablespoon butter

1. Dice potato and celery. Skin may be left on potato.
2. Sauté in skillet with butter and soy sauce until lightly
   browned.

3. Add sprouts and stir 1 minute.
4. Stuff hens.
5. Bake hens according to package directions. As they bake, brush from time to time with a glaze of 2 tablespoons soy sauce and 2 tablespoons melted butter.

YIELD: Stuffing for 2 hens.

SALADS

My Favorite Salad

    1 head lettuce, preferably romaine or Boston
    ½ pound fresh spinach
    1 to 2 cups sprout combination (lentil, mung, alfalfa)
    4 carrots, grated
    ½ cup raisins, plumped in 1 cup water and drained
    ½ cup hulled sunflower seeds
    ¾ to 1 cup vegetable oil
    6 tablespoons cider vinegar or lemon juice
    ¼ cup honey
    1 teaspoon celery seeds
    ½ teaspoon salt
       Freshly ground pepper

1. Wash lettuce and spinach well. Remove tough stems from spinach and bruised leaves from lettuce. Dry greens and tear into bowl.
2. Toss in sprouts, carrots, raisins and sunflower seeds.
3. Combine remaining ingredients to make dressing and shake well.
4. Mix dressing into salad.

YIELD: 6 to 8 servings.

Three Beans and Three Sprouts Salad

Grow sprouts simultaneously for this excellent picnic salad. Use fresh-cooked wax and green beans if you wish.

　　3 cups alfalfa, mung and lentil sprouts, sprouted in combination
　　1 can (1 pound) wax beans, drained
　　1 can (1 pound) green beans, drained
　　1 can (1 pound) kidney beans or garbanzos, drained and rinsed
　　¾ cup cider vinegar
　　½ cup salad oil
　　¾ cup sugar
　　2 tablespoons soy sauce
　　2 teaspoons dry mustard
　　1 cup chopped onion
　　　 Salt and pepper to taste

1. Put sprouts and beans in bowl.
2. Mix remaining ingredients and purée in blender.
3. Pour over beans and sprouts, toss well, cover and marinate in refrigerator overnight.

YIELD: 10 to 12 servings.

Hot Sprout Salad

　　3 slices bacon
　　2 cups whole green beans
　　2 cups sprouts
　　2 scallions, chopped
　　2 tablespoons vegetable oil
　　1 teaspoon flour
　　¼ cup cider vinegar

¼ cup warm water
Salt and pepper

1. Fry bacon crisp in skillet. Let drain on paper towels.
2. In 1 tablespoon of bacon fat heat drained beans 2 minutes. Place beans in casserole or other serving dish.
3. In 1 tablespoon oil heat sprouts and scallions 2 minutes. Stir into casserole.
4. Clean skillet with paper towel. Add 1 tablespoon oil. Heat. Stir in flour, vinegar and water. Add salt and pepper, stir until boiling. Pour sauce over sprouts and beans. Crumble bacon over top. Keep warm until ready to serve.

YIELD: 6 servings.

Molded Sprout Slaw

This salad works well with precooked soy, garbanzo, and pinto sprouts, or raw lentil, mung or wheat sprouts.

1 package lemon Jello
1⅔ cups hot water
The juice of 1 lemon
1 teaspoon salt
2 tablespoons chopped chives
1 cup shredded cabbage
½ cup shredded carrots
½ cup chopped celery
1 cup sprouts

1. Dissolve Jello in hot water.
2. Add lemon juice, salt and chives.
3. Chill until partially set.
4. Add vegetables and chill in mold until firm.
5. Serve with any mayonnaise dressing. Try mixing mayonnaise with a few drops lemon juice and curry powder to taste.

YIELD: 6 to 8 servings.

Noodle-Sprout Salad

A beautiful buffet showpiece.

        4 cups medium noodles
        7 tablespoons salad oil
        1 bunch watercress
        3 cups mung or lentil sprouts
        2 diced tomatoes
        ½ cup chopped walnuts or pecans
        ½ cup toasted wheat germ
        2 tablespoons toasted sesame seeds
        ¾ teaspoon salt
        6 tablespoons lemon juice

1. Cook noodles till tender according to package directions.
2. Drain and rinse with cold water. Chill.
3. Toss noodles with 1 tablespoon of the oil.
4. On a platter arrange noodles in outer ring, then a ring of watercress, then sprouts, then tomatoes, with the nuts in the center.
5. Combine wheat germ, sesame seeds, salt, lemon juice and remaining oil. Pour over salad and serve.

   YIELD: 4 to 6 servings.

Apple-Sprout Salad

        1 small head of cabbage, grated fine
        1 large apple, chopped
        1 cup alfalfa, clover or mustard sprouts
        4 tablespoons tarragon vinegar
        2 tablespoons sugar

¼ teaspoon salt
3 tablespoons sour cream

1. Combine cabbage, apple and sprouts
2. Combine vinegar, sugar, salt and sour cream.
3. Stir dressing into salad. Add more sour cream if desired.

YIELD: 4 to 6 servings.

Molded Fruit Salad

1 envelope unflavored gelatin
1 6-ounce can unsweetened pineapple juice
1 small (8¾ ounces) can sliced peaches
1 cup sprouts (mung, alfalfa, lentil)
½ cup shredded coconut
1 tablespoon mint jelly
1 teaspoon grated orange or lemon rind

1. Soften gelatin in pineapple juice.
2. Drain peaches and place syrup in saucepan. Heat with 1 cup water until just below boiling point.
3. Add gelatin and remaining ingredients. Stir well.
4. Pour into mold. Cool to room temperature.
5. Refrigerate until set.

YIELD: 8 servings.

Creamy Sprout Dressing

1 cup alfalfa sprouts
1 cup diced pineapple
¼ cup toasted sesame seeds

1. Combine all ingredients in blender for 1 minute.
2. Serve over fruit, gelatin salads, sliced tomatoes, cabbage or salad greens.

YIELD: Approximately 2 cups.

SOUPS

Sprout Egg Drop Soup

   1½ quarts seasoned chicken or beef broth
    2 cups beansprouts (mung, lentil, soy, garbanzo)
    3 eggs, well beaten
    3 tablespoons minced parsley

1. Heat broth.
2. Add sprouts and simmer 3 minutes.
3. Remove from heat and stir in eggs with fork.
4. Garnish with parsley.

YIELD: 2 quarts soup.

Pat's Soy Sprout and Sparerib Soup

This soup is economical and extremely nutritious.

   2 pounds spareribs
   8 cups water
   3 cups soy sprouts
     Salt
     Pepper

1. Cut the spareribs into bite-sized pieces with a heavy knife or cleaver, or have your butcher cut them for you.
2. Brown the spareribs quickly.

3. Boil the water and add the browned spareribs and the soy sprouts.
4. Cook until the sprouts and spareribs are tender, about 45 minutes. Season with salt and pepper to taste.

YIELD: 4 to 6 servings.

Lentil Sprout Soup

This is a very unusual and hearty soup.

  3 cups lentil sprouts
  ¼ pound Canadian bacon
  2 onions, coarsely chopped
  2 large carrots, chopped
 16 prunes, pitted
  2 tablespoons brown sugar
  2 tablespoons vinegar
  ¼ teaspoon thyme
    Salt to taste
  4 to 6 cups water

1. Combine all ingredients in 3 cups water.
2. Simmer in heavy pot 2 to 3 hours until thick. Add more water if necessary.

YIELD: 6 servings.

Asparagus-Chicken-Sprout Soup

This thick and creamy soup is a whole meal in itself.

  1 10¾-ounce can cream of chicken soup
  1 8½-ounce can cut asparagus spears
  ½ cup milk or half-and-half
  ½ cup sprouts (mung, lentil)
  1 tablespoon butter
  ¼ teaspoon basil
  1 egg

1. Heat chicken soup, asparagus and asparagus liquid, stirring in the milk or half-and-half.
2. Add sprouts and simmer 5 minutes.
3. Heat butter in 6″ skillet. Beat egg, adding basil. Cook egg without stirring until it is a flat pancake. Remove from heat and turn onto cutting board.
4. Dice egg small. Serve soup garnished with egg.

YIELD: 2 main-course servings, 4 soup-course servings.

Liver-Sprout Dumpling Soup

  2 slices rye bread
  ½ cup milk
  1 small onion
  ¾ cup just-budded sprouts (rye, wheat, lentil)
     8-ounce package liver sausage
  2 10¾-ounce cans chicken or beef broth
  1 cup water

1. Soak bread in milk for 10 minutes.
2. Grind onion and sprouts in meat grinder or blender.
3. Squeeze milk out of bread; discard milk.
4. Blend onion, sprouts, bread and sausage together.
5. Bring broth and 1 cup water to a boil.
6. Form liver mixture into 1″ balls and drop into rapidly boiling broth. Simmer 15 minutes and serve.

YIELD: 4 servings.

VEGETABLES

Sautéed Sprouts

This is the basic and most common way to use sprouts as a vegetable. When using soy, garbanzo, pea and some of the tougher beansprouts, you will want to steam them for 10 minutes first.

84

Vegetable oil
Garlic, onion, scallions or shallots
Sprouts
Broth, gravy or vegetable juice

1. Heat oil and sauté garlic or onion or other seasoning until tender.
2. Add sprouts and stir-fry quickly 3 minutes.
3. Moisten with broth, gravy or juice.
4. Season and serve. Good seasonings include soy sauce, salt, pepper, parsley, curry powder, herbs of your choosing.

Creole Sprouts

1 tablespoon vegetable oil
⅓ cup chopped onion
½ cup diced celery
1 can (1 pound) stewed tomatoes
1 bay leaf
½ teaspoon salt
2 cups sprouts (mung, soy, lentil or pea)

1. Heat oil in skillet, add onion and celery and sauté until golden brown.
2. Add tomatoes, bay leaf and salt, and bring to a boil. Simmer uncovered for 10 minutes.
3. Remove bay leaf. Add sprouts and simmer covered about 5 minutes, or longer if using a tough sprout.

YIELD: 4 servings.

Green Beans with Sprouts and Bacon

2 strips bacon
1 package frozen green beans
⅛ teaspoon curry powder
⅛ teaspoon chili powder
⅛ teaspoon garlic powder
¼ cup water
1 cup mung or lentil sprouts

1. Fry bacon in skillet until crisp. Remove bacon and drain.
2. Place partially thawed package of beans in hot bacon fat and stir to separate beans.
3. Season with curry, chili and garlic powders.
4. Add water, turn heat to low and simmer until the beans are tender.
5. Stir in sprouts and crumbled bacon just to heat them through.

YIELD: 4 to 6 servings.

Sprouts and Fresh Mushrooms

2 tablespoons butter
1 cup chopped onion
1 cup diced celery
1 cup sprouts—radish, alfalfa, clover or mustard
¾ cup fresh chopped mushrooms
Soy sauce
½ cup toasted almond slivers

1. Heat butter in skillet, add onion and celery and sauté 3 minutes.
2. Add sprouts and mushrooms, and steam, covered, about 5 minutes.
3. Add 1 to 2 tablespoons soy sauce. Remove from heat.
4. Garnish with almonds.

YIELD: 4 servings.

Vegetable Sukiyaki

2 tablespoons vegetable oil
1 pound fresh mushrooms, sliced
1 pound fresh spinach, washed and chopped
2 celery stalks, diced
1 cup mung or lentil sprouts
1 cup soy, garbanzo or pea sprouts, steamed
½ cup clover, mustard or radish sprouts
½ cup soy sauce
2 tablespoons Chinese brown sauce (or molasses)
¼ cup bouillon
Salt
Pepper

1. In a large skillet heat oil and brown mushrooms.
2. Add remaining vegetables and more oil, if necessary. Add soy sauce, brown sauce and bouillon.
3. Stir-fry 5 to 10 minutes until all vegetables are tender. Season to taste with salt and pepper.

YIELD: 6 servings.

Individual Sprout Soufflés

3 eggs
1½ cups sprouts, ground (soy, garbanzo, pea)
½ teaspoon salt
1 package powdered vegetable broth
1 tablespoon parsley
1 tablespoon chives
Butter
Bread crumbs
Tomato or mushroom sauce

1. Separate egg whites and yolks. Beat whites until stiff.
2. Beat yolks. Add sprouts, salt, vegetable broth, parsley, chives.
3. Fold in egg whites.

4. Butter custard cups well; dust with bread crumbs.
5. Fill cups ⅔ full. Set in pan of hot water.
6. Bake at 350° F. 25 to 35 minutes until puffed and brown. Serve with sauce.

YIELD: 6 soufflés.

DESSERTS, CANDIES, SNACKS

Soybean Snacks

This is a nutritious, delicious snack that keeps well for weeks. Carry it in the car, to school, to work. Serve in small bowls in place of nuts.

Method I
This makes the sprouts taste like fresh-roasted peanuts.

    2 cups soybean sprouts (or garbanzo or pea sprouts)
      Vegetable oil.
      Salt

1. Put oil in deep saucepan to the depth of 1 inch. Heat to about 350° F.
2. Rinse and drain sprouts thoroughly. Pat dry between layers of absorbent towels.

3. Deep-fry sprouts, a few at a time. Caution: fat will bubble up.
4. Remove when they are golden brown and drain on paper towels.
5. Salt to taste.

Method II
Good if you are avoiding fats in your diet.

1. Place sprouts in single layer in a baking pan.
2. Bake at 350° F. about ½ hour, until sprouts are golden brown.
3. Salt to taste.

Cream Cheese Rolls

   1 cup sprouts (wheat, rye, alfalfa)
   4 ounces cream cheese
   1 cup raisins
   1 cup nuts or dried chopped sprouts or coconut

1. Blend first 3 ingredients together.
2. Roll into balls.
3. Coat with nuts, sprouts or coconut.

YIELD: 2 dozen rolls.

Dried Sprout Candy

    4 cups dried sprouts (alfalfa, wheat, rye)
    ¼ cup nutritional yeast
    ½ cup powdered milk
    1 cup peanut butter
    1 cup dried currants or raisins
    ½ cup honey

1. Grind sprouts in blender.
2. Mix all ingredients together.
3. Roll into balls.
4. Roll balls into topping if desired (nuts, coconut, sesame or sunflower seeds, carob or chocolate).
5. Chill.

    YIELD: 4 to 5 dozen candies.

Sprout Pie

This recipe is hard to believe in, so don't tell anyone that they are eating a vegetable for dessert. It is super-nutritious and delicious, with a crunchy nut texture and a flavor similar to that of pumpkin pie.

    ¾ cup dark-brown sugar
    1 13-ounce can evaporated milk
    ½ teaspoon salt
    1 teaspoon cinnamon
    ½ teaspoon nutmeg
    ¾ teaspoon powdered ginger
    1½ cups soybean sprouts, ground
    2 eggs, lightly beaten
      Unbaked pie crust

1. Combine sugar, ⅓ of the milk, salt, cinnamon, nutmeg and ginger.
2. Put sprouts through meat grinder. Pack down firmly to measure.
3. Stir sugar mixture, sprouts and eggs together very well, slowly adding rest of the milk.

4. Pour into pie shell and bake at 350° F. about 45 minutes until knife inserted in center comes out clean.
5. Serve with ice cream or whipped cream.

YIELD: One 9″ pie.

Sprout Lemon Tea Cookies

   2¾ cups flour
   ½ teaspoon salt
   4 teaspoons baking powder
   ⅔ cup shortening
 1⅔ cups sugar
   2 eggs, beaten
   2 teaspoons lemon flavoring or juice
 2¾ cups ground sprouts (soy, garbanzo)
   ⅓ cup milk

1. Sift flour, salt and baking powder.
2. Cream shortening with sugar; add eggs, lemon flavoring and sprouts.
3. Add milk and dry ingredients alternately. Mix well.
4. Drop by teaspoonfuls onto an oiled baking sheet and bake in a 375° F. oven about 20 minutes, until lightly brown at edges.

YIELD: About 5 dozen medium cookies.

Oatmeal Sprout Cookies

   1 cup vegetable oil
   1 cup brown sugar
   2 eggs
   4 tablespoons milk
   1 teaspoon vanilla
   ½ cup flour
   2 cups ground toasted sprouts (soy, garbanzo)
 2½ cups raw oatmeal

1. Cream oil with brown sugar. Add eggs, milk and vanilla.
2. Fold in flour, sprouts and oatmeal. Mix well.
3. Drop by teaspoonfuls onto oiled cookie sheet.
4. Flatten with fork dipped into cold water.
5. Bake at 350° F. until light brown, about 10 minutes.

YIELD: About 3 dozen.

75 SPROUT IDEAS

Here's a list to inspire you to experiment and improvise with sprouts in *your* own cooking. The sprouts in parenthesis work well, but try others too.

1. Blend into milk shakes (almond, soy, sesame).
2. Steep in water to make sprout tea (alfalfa, fenugreek).
3. Blend with apple or pineapple juice (alfalfa, clover).
4. Add to vegetable juice (soy, mung, lentil).
5. Use to replace celery in sandwich spreads (radish, lentil).
6. Add to grilled cheese sandwiches (clover, alfalfa).
7. Grind and use as a sandwich spread (soy, garbanzo).
8. Blend with soft cheeses as a dip (sesame, mung).
9. Decorate canapés (lentil, mung, mustard).
10. Marinate and serve as an hors d'oeuvre (soy, lentil).
11. Add to tossed salads (all sprouts).
12. Use in coleslaw (mung, clover, radish).
13. Use in potato salad (lentil, soy, mung).
14. Add to egg filling for stuffed tomatoes (mung).
15. Add to aspics (mung, soy, garbanzo).
16. Use in jellied fruit salads (alfalfa, rye).
17. Sprinkle on fruit yogurt (alfalfa).
18. Stir into soups at the last minute (rye, lentil).
19. Cook with vegetable soups (soy, pea).
20. Replace part of cooked cereal (oat, wheat).
21. Sprinkle on dry cereal (alfalfa, wheat, rye).

22. Toast and eat as a snack (soy, pea).
23. Stir into pancake or waffle batter (buckwheat).
24. Combine with rice dishes (fenugreek, rye).
25. Add to fried rice (mung, soy, lentil).
26. Stir into spaghetti sauce (clover, mustard).
27. Use in lasagne (mung, lentil).
28. Mix with cheese for stuffed ravioli (lentil).
29. Add to vegetable soufflés (clover, mustard).
30. Use in vegetable fritters (wheat, alfalfa).
31. Add to turkey or beef pot pie (lentil, pea).
32. Stir into chicken or ham à la king (mung, soy).
33. Add to scrambled eggs (soy, sesame).
34. Make a sprout herb omelet (clover, radish).
35. Stew with tomatoes (mung, lentil, soy).
36. Sauté with onions (garbanzo, mung).
37. Purée with peas or beans (soy, pea, mung).
38. Stuff mushroom caps and broil (mustard, alfalfa).
39. Substitute for Chinese vegetables (mung, lentil).
40. Add to baked beans (soy, lentil).
41. Add to vegetarian meat-replacement dishes (soy).
42. Braise with celery (mung, lentil, pea).
43. Steam and serve with butter (pea, soy, garbanzo).
44. Stuff with corn into peppers (mung, lentil).
45. Grind and mash into potatoes (rye, soy).
46. Add to potato pancakes (alfalfa).
47. Stuff squash (lentil, radish).
48. Mix with hashed brown potatoes (lentil).
49. Sprinkle on tomatoes and broil (mustard, clover).
50. Purée and add to salad dressings (soy, mung).
51. Blend with mayonnaise for spreading (lentil, pea).
52. Use as poultry or fish stuffing (mung, soy).
53. Add as a crunchy topping to casseroles (clover).
54. Stir into pickle relish (garbanzo, pea).
55. Line oyster shells, top with creamed oysters (mung).
56. Stir into shrimp creole (mung, lentil).
57. Add to any Chinese or Japanese dish (mung, soy).
58. Use as a bed for fish (alfalfa, clover).
59. Stir into stews at the last moment (soy, mung).

60. Stuff an eggplant (soy, lentil).
61. Grind and add to meat, fish or nut loaves (soy).
62. Garnish a platter (lentil, alfalfa, mung).
63. Replace celery in poultry stuffing (mung, lentil).
64. Pickle in brine (soy, garbanzo, pea).
65. Use in stuffed cabbage or grape leaves (mung).
66. Add to breads (wheat, rye).
67. Stir into fruit and nut bars (rye, wheat).
68. Bake in fruit charlottes (alfalfa).
69. Add to whole wheat breads (wheat, rye).
70. Add to nut cookies, peanut butter cookies (alfalfa, rye).
71. Use in steamed puddings (alfalfa, rye).
72. Stir into nut candy batters (alfalfa, rye).
73. Cook with fruit cakes (alfalfa, wheat).
74. Add to dried camping foods as you cook them outdoors (soy).
75. And—why not Sprout Ice Cream!

List of Suppliers

Since most mail-order suppliers charge postage or freight, you will want to order from a reliable source as near your home as possible. Remember that dried seeds and beans are heavy. To locate a supplier in your vicinity, or a local health food store, you may want to order *The Organic Directory* (Rodale Press, Emmaus, Pennsylvania 18049, $1.95).

Beale's Famous Products, Box 323, Ft. Washington, Pa. 19034. Specialists in sprout products. Unglazed pottery sprouter with directions and sample seeds, excellent assortment of 15 different seeds for testing your favorites, breakfast mix, salad mix, sandwich mix, books. Shipping extra. Send for price lists.

Riggscraft, P.O. Box 1273, Laramie, Wyoming 82070. Shippers of sprout-quality seeds, beans and grains. Service and shipping charges extra. Very reasonable. Free catalogue.

Walnut Acres, Penns Creek, Pa. 17862. Almost every sprout product: seeds, beans, grains, sprouters and supplies; stainless steel mesh for jars. Reasonable. Shipping extra. Free catalogue.

Natural Development Co., Bainbridge, Pa. 17502. Cress seeds, sunflower seeds, mung beans, buckwheat, lentils, corn, wheat, alfalfa, soybeans. Very reasonable. Free postage east of the Mississippi River. Free catalogue.

Shiloh Farms, Rt. 59, Sulphur Springs, Ark. 72768. Ask for price lists.

Aphrodisia, 28 Carmine St., New York, N.Y. 10014. Alfalfa, fenugreek, flax, black and yellow mustard seeds, safflower, sesame seeds. Shipping extra. Catalogue.

Brownie's Natural Foods, 21 E. 16th St., New York, N.Y. 10003. Catalogue.

Kwong On Long Co., 686 North Spring St., Los Angeles, Calif. 90012. Mail order soy and mung beans, other ingredients for oriental cooking.

Wing Fat Co., 35 Mott St., New York, N.Y. 10013. Mung beans, soybeans, and other ingredients for Oriental cooking.

Barth's of Long Island, 270 W. Merrick Road, Valley Stream, N.Y. 11582.

Bibliography

Chen, Philip S. *Soybeans: For Health, Longevity, and Economy*. Provoker Press, St. Catherine's, Ont., Canada, 1970.

Dinshah, Freya. *The Vegan Kitchen*. American Vegan Society, Malaga, N.J., 1969.

El Molino Mills. *Best Recipes*. Alhambra, Calif., 1953.

Elwood, Catharyn. *Feel Like a Million*. Pocket Books, New York, 1969.

Hiatt, D.R. *The Sprouting of Fresh Seed for Food*. Message Press, Coalmont, Tenn.

Hunter, Beatrice Trum. *The Natural Foods Cookbook*. Simon and Schuster, New York, 1961.

Jensen, Dr. Bernard. *Seeds and Sprouts for Life*. Hidden Valley Health Ranch, Escondido, Calif.

Jones, Dorothea Van Gundy. *The Soybean Cookbook*. Arc, New York, 1971.

Keys, Margaret and Ancel, *The Benevolent Bean*. Noonday, N.Y., 1967.

Kulvinskas, Viktoras. *Love Your Body*. Happy Souls, Wethersfield, Connecticut, 1972.

Larsen, Gena. *Fundamentals in Foods*. International Association of Cancer Victims and Friends, Inc., San Diego, Calif., 1967.

The Organic Directory. Rodale Press, Emmaus, Pa., 1971.

Tobe, John H. *Sprouts, Elixir of Life*, Modern Publication, St. Catharines, Ont., Canada, 1968.

U.S. Department of Agriculture. *Composition of Foods, Agriculture Handbook #8*. U.S. Government Printing Office, Washington, D.C., 1963.

Wheelwright, A. Stuart. *Sprout Handbook*. Research Technical Service, Ogden, Utah, 1968.